D0399277

DEEP THINKING

DEEP THINKING

where machine intelligence ends and human creativity begins

GARRY KASPAROV

WITH MIG GREENGARD

PUBLICAFFAIRS
New York

Published by PublicAffairs™, an imprint of Perseus Books, LLC,
a subsidiary of Hachette Book Group, Inc.

Printed in the United States of America.

The Hachette Speakers Bureau provides a wide range of authors for speaking events. To find out more, go to hachettespeakersbureau.com or call 866-376-6591.

Library of Congress Cataloging-in-Publication data is available for this book.

ISBN 978-1-61039-786-5 (HC)
ISBN 978-1-61039-787-2 (EB)

Editorial production by Christine Marra, *Marra*thon Production Services.
www.marrathoneditorial.org

Book design by Jane Raese
Set in 10-point Utopia

FIRST EDITION

10 9 8 7 6 5 4 3 2 1

FOR MY CHILDREN, POLINA, VADIM, AIDA, AND NICKOLAS.

Challenge yourselves and you will challenge the world.

CONTENTS

DEEP THINKING

INTRODUCTION

IT WAS A PLEASANT DAY in Hamburg on June 6, 1985, but chess players rarely get to enjoy the weather. I was inside a cramped auditorium, pacing around inside a circle of tables upon which rested thirty-two chessboards. Across from me at every board was an opponent, who moved promptly when I arrived at the board in what is known as a simultaneous exhibition. "Simuls," as they are known, have been a staple of chess for centuries, a way for amateurs to challenge a champion, but this one was unique. Each of my opponents, all thirty-two of them, was a computer.

I walked from one machine to the next, making my moves over a period of more than five hours. The four leading chess computer manufacturers had sent their top models, including eight bearing the "Kasparov" brand name from the electronics firm Saitek. One of the organizers warned me that playing against machines was different because they would never get tired or resign in dejection the way a human opponent would; they would play to the bitter end. But I relished this interesting new challenge—and the media attention it attracted. I was twenty-two years old, and by the end of the year I would become the youngest world chess champion in history. I was fearless, and, in this case, my confidence was fully justified.

It illustrates the state of computer chess at the time that it didn't come as much of a surprise, at least not in the chess world, when I achieved a perfect 32–0 score, winning every game, although there was one uncomfortable moment. At one point I realized that I was drifting into trouble in a game against one of the Kasparov models. If this machine scored a win or even a draw against me, people might suggest that I had thrown the game to get publicity for the company, so I had to intensify my efforts. Eventually I found a way to trick the machine with a sacrifice it should have refused and secure my clean sweep. From the human perspective, or at least from my perspective as the human in this equation, these were the good old days of human versus machine chess. But this golden age would be brutally short.

Twelve years later I was in New York City fighting for my chess life against just one machine, a $10 million IBM supercomputer nicknamed "Deep Blue." This battle, actually a rematch, became the most famous human-machine competition in history. *Newsweek*'s cover called the it "The Brain's Last Stand" and a flurry of books compared it to Orville Wright's first flight and the moon landing. Hyperbole, of course, but not out of place at all in the history of our love-hate relationship with so-called intelligent machines.

Jump forward another twenty years to today, to 2017, and you can download any number of free chess apps for your phone that rival any human Grandmaster. You can easily imagine a robot in my place in Hamburg, circling inside the tables and defeating thirty-two of the world's best human players at the same time. The tables have turned, as they always do in our eternal race with our own technology.

Ironically, if a machine did perform a chess simul against a room full of human professional players, it would have more trouble moving from board to board and physically moving the pieces than it would have calculating the moves. Despite centuries of science fiction about automatons that look and move like people, and for all the physical labor today done by robots, it's fair to say that we have advanced further in duplicating human thought than human movement.

In what artificial intelligence and robotics experts call Moravec's paradox, in chess, as in so many things, what machines are good at is where humans are weak, and vice versa. In 1988, the roboticist Hans Moravec wrote, "It is comparatively easy to make computers exhibit adult level performance on intelligence tests or playing checkers, and difficult or impossible to give them the skills of a one-year-old when it comes to perception and mobility." I wasn't aware of these theories at the time, and in 1988 it was safe to include checkers but not yet chess, but ten years later it was obviously the case in chess as well. Grandmasters excelled at recognizing patterns and strategic planning, both weaknesses in chess machines that, however, could calculate in seconds tactical complications that would take even the strongest humans days of study to work out.

This disparity gave me an idea for an experiment after my matches with Deep Blue attracted so much attention. You could also call it "if you can't beat 'em, join 'em," but I was eager to continue the computer chess experiment even if IBM was not. I wondered, what if instead of human versus machine we played as partners? My brainchild saw the light of day in a match in 1998 in León, Spain, and we called it Advanced Chess. Each player had a PC at hand running the chess software of his choice during the game. The idea was to create the highest level of chess ever played, a synthesis of the best of man and machine. It didn't quite go according to plan, as we'll see later, but the fascinating results of these "centaur" competitions convinced me that chess still had a lot to offer the worlds of human cognition and artificial intelligence.

In this belief I was hardly a pioneer; a chess-playing machine has been a holy grail since long before it was possible to make one. I just happened to be the human holding the grail when it was finally in science's grasp. I could run away from this new challenge or I could embrace it, which was really no choice at all. How could I resist? It was a chance to promote chess to a general audience beyond that reached even by Bobby Fischer's Cold War–era match against Boris Spassky and my own title duels with Anatoly Karpov. It had the potential to attract a new set of deep-pocketed sponsors to chess, especially tech companies. For example, Intel sponsored a Grand Prix cycle in the mid-1990s as well as my world championship match with Viswanathan Anand in 1995, played at the top of the World Trade Center. And then there was the irresistible curiosity I felt. Could these machines really play chess at the world championship level? Could they really think?

Humans have dreamed of intelligent machines since long before the technology to attempt one was conceived. In the late eighteenth century, a chess-playing mechanical automaton called the "Turk" was a wonder of the age. A carved wooden figure moved the pieces and, most remarkably, played a very strong game. Before it was destroyed in a fire in 1854, the Turk toured Europe and the Americas to great acclaim, claiming among its victims the famous chess aficionados Napoleon Bonaparte and Benjamin Franklin.

Of course it was a hoax; there was a human inside the cabinet under the table, hidden by an ingenious set of sliding panels and machinery. In another irony, today chess tournaments are plagued by cheaters who access super-strong computer programs to defeat their human opponents. Players have been caught using sophisticated signaling methods with accomplices, Bluetooth headsets in hats or electrical devices in shoes, and simply using a smartphone in the restroom.

The first real chess program actually predates the invention of the computer and was written by no less a luminary than Alan Turing, the British genius who cracked the Nazi Enigma code. In 1952, he processed a chess algorithm on slips of paper, playing the role of CPU himself, and this "paper machine" played a competent game. This connection went beyond Turing's personal interest in chess. Chess had a long-standing reputation as a unique nexus of the human intellect, and building a machine that could beat the world champion would mean building a truly intelligent machine.

Turing's name is forever attached to a thought experiment later made real, the "Turing test." The essence is whether or not a computer can fool a human into thinking it is human and if yes, it is said to have passed the Turing test. Even before I faced Deep Blue, computers were beginning to pass what we can call the "chess Turing test." They still played poorly and often made distinctively inhuman moves, but there were complete games between computers that wouldn't have looked out of place in any strong human tournament. As became clearer as the machines grew stronger every year, however, this taught us more about the limitations of chess than about artificial intelligence.

You cannot call the globally celebrated culmination of a forty-five-year-long quest an anticlimax, but it turned out that making a great chess-playing computer was not the same as making a thinking machine on par with the human mind, as Turing and others had dreamed.

Deep Blue was intelligent the way your programmable alarm clock is intelligent. Not that losing to a $10 million alarm clock made me feel any better.

The AI crowd, too, was pleased with the result and the attention but dismayed by the fact that Deep Blue was hardly what their predecessors had imagined decades earlier when they dreamed of creating a machine to defeat the world chess champion. Instead of a computer that thought and played chess like a human, with human creativity and intuition, they got one that played like a machine, systematically evaluating up to 200 million possible moves on the chess board per second and winning with brute number-crunching force. This isn't to diminish the achievement in any way. It was a human achievement, after all, so while a human lost the match, humans also won.

After the unbearable tension of the match, exacerbated by IBM's questionable behavior and my suspicious human mind, I was in no mood to be a gracious loser. Not that I've ever been a good loser, I hasten to add. I believe accepting losses too easily is incompatible with being a great champion—certainly this was the case with me. I do believe in fighting a fair fight, however, and this is where I felt IBM had shortchanged me as well as the watching world.

Reexamining every aspect of that infamous match with Deep Blue for the first time in twenty years has been difficult, I admit. For two decades I have succeeded almost completely in avoiding and deflecting discussion about my Deep Blue matches beyond what was publicly known. There are many books about Deep Blue, but this is the first one that has all the facts and the only one that has my side of the story. Painful memories aside, it has also been a revealing and rewarding experience. My great teacher Mikhail Botvinnik, the sixth world champion, taught me always to seek the truth in the heart of every position. It has been fulfilling to finally find the truth at the heart of Deep Blue.

MY CAREER and my investigations into human-machine cognition did not end with Deep Blue, however; nor does this book. In fact, in both cases it's just the beginning. Competing head to head against a computer the way I did isn't the norm, although it was symbolic of how we are in a strange competition both with and against our own creations in more ways every day. My Advanced Chess experiment flourished online, where teams of humans and computers working together competed with remarkable results. Smarter computers are one key to success, but doing a smarter job of humans and machines working together turns out to be far more important.

These investigations led to visits to places like Google, Facebook, and Palantir, companies for whom algorithms are lifeblood. There have also been some more surprising invitations, including one from the headquarters of the world's largest hedge fund, where algorithms make or lose billions of dollars every day. There I met one of the creators of Watson, the *Jeopardy*-playing computer that could be called IBM's successor to Deep Blue. Another trip was to participate in a debate in front of an executive banking audience in Australia on what impact AI was likely to have on jobs in their industry. Their interests are quite different, but they all want to be on the cutting edge of the machine intelligence revolution, or at least to not be cut by it.

I've been speaking to business audiences for many years, usually on subjects like strategy and how to improve the decision-making process. But in recent years, I'm receiving more and more requests to talk about artificial intelligence and what I call the human-machine relationship. Along with sharing my thoughts, these appearances have given me the opportunity to listen closely to the interests of the business world regarding intelligent machines. Much of this book is dedicated to addressing these concerns and separating inevitable facts from conjecture and hyperbole.

In 2013, I was honored to become a senior visiting fellow at the Oxford Martin School, where I get to spend time with a constellation of brilliant expert minds. At Oxford, artificial intelligence is as much an area of philosophy as technology, and I enjoy trying to cross these streams. Their wonderfully named Future of Humanity Institute is the perfect place to collaborate on where the human-machine relationship is headed. My goal is to take some of the sophisticated, often arcane, expert research, predictions, and opinions and to serve as your translator and guide to their practical implications while adding my own insights and questions along the way.

I have spent most of my life thinking about how humans think and have found this to be an excellent basis for relating how machines think, and how they do not. In turn, this insight helps inform us as to what our machines can and cannot do . . . yet.

THE NINETEENTH-CENTURY African American folk legend of John Henry pits the "steel-driving man" in a race against a new invention, a steam-powered hammer, bashing a tunnel through a mountain of rock. It was my blessing and my curse to be the John Henry of chess and artificial intelligence, as chess computers went from laughably weak to nearly unbeatable during my twenty years as the world's top chess player.

As we will see, this is a pattern that has repeated over and over for centuries. People scoffed at every feeble attempt to substitute clumsy, fragile machines for the power of horses and oxen. We laughed at the idea that stiff wood and metal could replicate the soaring grace of the birds. Eventually we have had to concede that there is no physical labor that couldn't be replicated, or mechanically surpassed.

It is also now widely accepted that this inexorable advance is something to celebrate, not fear, although it is usually two steps forward and one step back in this regard. With every new encroachment of machines, the voices of panic and doubt are heard, and they are only getting louder today. This is partly due to the differences in what, and who, is being replaced. The horses and oxen couldn't write letters to the editor when cars and tractors came along. Unskilled laborers also lacked much of a voice, and were often considered lucky to be freed from their backbreaking toil.

So it went over the decades of the twentieth century, with countless jobs lost or transformed by automation. Entire professions disappeared with little time to mourn them. The elevator operators' union was seventeen thousand strong in 1920, although its ability to paralyze cities with strikes like the one its members staged in New York in September 1945 surely cost them more than a few mourners when automatic push-button elevators began to replace them in the 1950s. According to the Associated Press, "Thousands struggled up stairways that seemed endless, including the Empire State Building, tallest structure in the world."

Good riddance, you might imagine. But the worries about operatorless elevators were quite similar to the concerns we hear today about driverless cars. In fact, I learned something surprising when I was invited to speak to the Otis Elevator Company in Connecticut in 2006. The technology for automatic elevators had existed since 1900, but people were too uncomfortable to ride in one without an operator. It took the 1945 strike and a huge industry PR push to change people's minds, a process that is already repeating with driverless cars. The cycle of automation, fear, and eventual acceptance goes on.

Of course, what an observer calls freedom and disruption, a worker calls unemployment. The educated classes in the developed world have long had the luxury of lecturing their blue-collar brethren about the glories of the automated future. Service personnel have been on the block for decades—their friendly faces, human voices, and quick fingers replaced by ATMs, photocopiers, phone trees, and self-checkout lines. Airports have iPads instead of food servers. No sooner did massive call centers spring up around India than automated help-desk algorithms begin replacing them.

It is far easier to tell millions of newly redundant workers to "retrain for the information age" or to "join the creative entrepreneurial economy" than to be one of them or to actually do it. And who can say how quickly all that new training will also become worthless? What professions today can be called "computer proof"? Today another set of tables has turned, or rather, desks. The machines have finally come for the white collared, the college graduates, the decision makers. And it's about time.

JOHN HENRY won his race against the machine only to die on the spot, "his hammer in his hand." I was spared such a fate myself, and humans are still playing chess, in fact more today than ever before. The doomsayers who said no one would want to play a game that could be dominated by a computer have been proven wrong. This seems obvious, considering how we also still play far simpler games like tic-tac-toe and checkers, but doomsaying has always been a popular pastime when it comes to new technology.

I remain an optimist if only because I've never found much advantage in the alternatives. Artificial intelligence is on a path toward transforming every part of our lives in a way not seen since the creation of the Internet, perhaps even since we harnessed electricity. There are potential dangers with any powerful new technology and I won't shy away from discussing them. Eminent individuals from Stephen Hawking to Elon Musk have expressed their fear of AI as a potential existential threat to mankind. The experts are less prone to alarming statements, but they are quite worried too. If you program a machine, you know what it's capable of. If the machine is programming itself, who knows what it might do?

The airports with their self-check-in kiosks and restaurants full of iPads are staffed by thousands of human workers (most using machines) in the long security lines. Is it because they can do things no machine can do? Or, like operating an elevator and driving a car, is it because at first we don't trust machines to do a job where lives are at risk? Elevators became much safer as soon as the human operators were replaced. The human-hating Skynet from the *Terminator* movies could hardly do a better job of killing people than we do killing ourselves with cars. Human error is responsible for over 50 percent of plane crashes, although overall air travel is getting safer as it becomes more automated.

In other words, fail-safes are required, but so is courage. When I sat across from Deep Blue twenty years ago I sensed something new, something unsettling. Perhaps you will experience a similar feeling the first time you ride in a driverless car, or the first time your new computer boss issues an order at work. We must face these fears in order to get the most out of our technology and to get the most out of ourselves.

Many of the most promising jobs today didn't even exist twenty years ago, a trend that will continue and accelerate. Mobile app designer, 3D print engineer, drone pilot, social media manager, genetic counselor—to name just a few of the careers that have appeared in recent years. And while experts will always be in demand, more intelligent machines are continually lowering the bar to creating with new technology. This means less training and retraining for those whose jobs are taken by robots, a virtuous cycle of freeing us from routine work and empowering us to use new technology productively.

Machines that replace physical labor have allowed us to focus more on what makes us human: our minds. Intelligent machines will continue that process, taking over the more menial aspects of cognition and elevating our mental lives toward creativity, curiosity, beauty, and joy. These are what truly make us human, not any particular activity or skill like swinging a hammer—or even playing chess.

CHAPTER 1

THE BRAIN GAME

CHESS IS OLD ENOUGH for its origins to be less than entirely clear. Most histories place the origins of the chess precursor game chaturanga in India sometime before the sixth century. From there chess moved to Persia and into the Arab and Muslim world, where it followed the well-trod path into southern Europe via Moorish Spain. By the time of the late Middle Ages, it was a standard presence in the courts of Europe and appears regularly in manuscripts from the period.

The modern game we know today appeared in Europe at the end of the fifteenth century, when the ranges of the queen and bishop were extended, making the game far more dynamic. Older and regional variants still existed, and there were a few minor rule standardizations, but for the most part, games played by the eighteenth century were identical to those played today. This rich history includes thousands of games from great masters of centuries past, with each move, each brilliancy and each blunder, perfectly preserved in chess notation as if trapped in amber.

The games are what matter most to serious players, but history and physical relics also play a role in the game's status. The twelfth-century Lewis chessmen, carved from walrus tusks; illuminated Persian illustrations from 1500 of players accompany Rumi's poetry; the third book ever printed in English was *Game and Playe of the Chesse,* which came from the press of William Caxton himself in 1474; Napoleon Bonaparte's personal chess set. You start to see why chess fans resent it being called just a game.

This global heritage is what makes chess unique as a cultural artifact, but the fact of its longevity and popularity doesn't explain it. The number of people who play chess regularly is impossible to know exactly, of course, but some of the more extensive surveys with modern sampling methods put the figure in the hundreds of millions. The game is popular on every continent, with regional concentrations from its traditional popularity in the former Soviet and Soviet Bloc countries and from its recent boom in India, which is due largely to the successes of former world champion Viswanathan Anand.

My personal and entirely unscientific survey method is based on how often I am recognized in public when I travel, which I do most of the year. In the United States, where I now live in New York City, I can pass in anonymity for days at a time before being recognized, often by someone from Eastern Europe. For better or worse, chess champions can safely walk the streets of America without worrying about autograph hounds and paparazzi. Meanwhile, I was so mobbed by chess fans at my hotel during a lecture trip to New Delhi that the hotel had to have security escort me through, so I can't even imagine what it's like there for their national idol Anand.

The Soviet heyday, when chess champions were met by cheering crowds at train stations and airports, survives today only in chess-mad Armenia, where the national team has brought home gold medals at an astounding rate for a country with a population of only 3 million people. And despite my own half-Armenian heritage, there is no genetic explanation necessary for this success. When a society emphasizes something, by custom or by mandate, results will follow, whether it's a state religion, a traditional art form, or chess.

Does the "why chess?" question find an answer in anything intrinsic to the game itself? Is there something uniquely attractive to chess's blend of strategic and tactical elements, its balance of preparation, inspiration, and determination? To be honest, I don't think so. It's true that the game has had the benefit of centuries of evolution, adapting to its surroundings like one of Darwin's finches. For example, the romantic Renaissance players made the game far more lively, accelerating the game just as the world of ideas accelerated around it. And who is to say that the eight-by-eight chessboard isn't somehow more pleasing or accessible to the human mind than the nine-by-nine shogi board or the fathomless nineteen-by-nineteen grid of Go stones? It's a diverting thought, but we don't really have to look much further than how the increasingly interconnected world of the Enlightenment led to the standardization of everything from spelling to beer recipes to chess rules. Had a ten-by-ten board been in vogue around 1750 that's probably what we'd be playing with today.

THE ABILITY to play chess well has always had a special mystique as a representation of intelligence, a statement that applies equally to both human and machine players. As a young chess star and world champion I personally experienced this mystique and its side effects more than just about anyone. For every truth around elite chess players—we do have good memories and concentration skills—there are at least a dozen misconceptions, both positive and negative.

Connections between chess skill and general intelligence are weak at best. There is no more truth to the thought that all chess players are geniuses than in saying that all geniuses play chess. In fact, one of the things that makes chess so interesting is that it's still unclear exactly what separates good chess players from great ones. Recently, sophisticated brain scans have started to illuminate which functions of the brain strong players rely on most, although psychologists have analyzed the matter extensively for decades with batteries of tests.

The results of all these investigations have so far confirmed the ineffable nature of human chess. The start of the game, called the opening phase, is mostly a matter of study and recall for professionals. We select openings from our personal mental library according to our preferences and preparation for our opponent. Move generation seems to involve more visuospatial brain activity than the sort of calculation that goes into solving math problems. That is, we literally visualize the moves and positions, although not in a pictorial way, as many early researchers assumed. The stronger the player, the more they demonstrate superior pattern recognition and doing the sort of "packaging" of information for recall that experts call "chunking."

Then comes understanding and assessing what we see in our mind's eye, the evaluation aspect. Different players of equal strength often have very different opinions of a given position and recommend entirely different moves and strategies. There is ample room here for disparate styles, creativity, brilliancy, and, of course, terrible mistakes. All this visualization and evaluation must be verified by calculation, the "I go here, he goes there, I go there" mechanics that novices rely on—and that many assume incorrectly to be what chess is all about.

Finally, the executive process must decide on a course of action, and it must decide *when* to decide. Time is limited in a serious game of chess, so how much of it do you use on a given move? Ten seconds or thirty minutes? Your clock is ticking and your heart is racing!

All these things are happening at once during every second of a chess game, which can last for six or seven stressful hours at the competitive level. Unlike machines, we also have to cope with emotional and physical responses during every moment, everything from worry and excitement about the position to tiredness, hunger, and the limitless distractions about everyday life that constantly float through our consciousness.

A character of Goethe's called chess a "touchstone of the intellect," while Soviet encyclopedias defined chess as an art, a science, and a sport. Marcel Duchamp, himself a strong player, said that "I have come to the personal conclusion that while all artists are not chess players, all chess players are artists." Brain scans will continue to better define exactly what goes on in the human brain during a chess game, and may even come to some conclusions about what makes one person a naturally superior player. But I remain confident that we will continue to enjoy chess, and to revere it, as long as we enjoy art, science, and competition.

Thanks to the Internet's matchless ability to spread myths and rumors, I've found myself bombarded with all sorts of misinformation about my own intellect. Spurious lists of "highest IQs in history" might find me between Albert Einstein and Stephen Hawking, both of whom have probably taken as many proper IQ tests as I have: zero. In 1987, the German news magazine *Der Spiegel* sent a small group of experts to a hotel in Baku to administer a battery of tests to measure my brainpower in different ways, some specially designed to test my memory and pattern recognition abilities.

I have no idea how closely these approximated a formal IQ test, nor do I much care. The chess tests proved I was very good at chess, the memory tests that I had a very good memory, neither of which was much of a revelation. My weakness, they told me, was "figural thinking," apparently proven after I blanked out for a while when tasked with filling in some dots with pencil lines. I have no idea what was, or wasn't, going through my mind at the time, but I have always had difficulty motivating myself to perform tasks I cannot see the point of, a tendency I now see reflected in my daughter Aida when it's time to do her homework.

When *Der Spiegel* asked me what I thought separated me, the world champion, from other strong chess players, I answered, "The willingness to take on new challenges," the same answer I would give today. The willingness to keep trying new things—different methods, uncomfortable tasks—when you are already an expert at something is what separates good from great. Focusing on your strengths is required for peak performance, but improving your weaknesses has the potential for the greatest gains. This is true for athletes, executives, and entire companies. Leaving your comfort zone involves risk, however, and when you are already doing well the temptation to stick with the status quo can be overwhelming, leading to stagnation.

AS FLATTERING AS all the "genius" mythmaking might sound, it's really more a case of flattery of chess itself. It is a perpetuation of hundreds of years of praise of chess masters as virtuosos and prodigies. In 1782, the great French player François-André Danican Philidor played two games simultaneously while blindfolded and was acclaimed as an intellect without parallel. As one contemporary newspaper account described it, "a phenomenon in the history of man and so should be hoarded among the best samples of human memory, till memory shall be no more." Flattering, but as good as Philidor was for his era, playing two games without sight of the board is easily in range of any competent player with a little practice. And while there have been various claims to the world record for simultaneous blindfold play, the modern official record is forty-six, set by a German player of average master strength.

Regardless of the origins, there is no doubt that chess is an enduring symbol of intellectual prowess and strategic thinking, as well as an overly popular metaphor for everything from politics to war to every kind of sport and even to romantic entanglements. Perhaps chess players should receive a commission every time a football coach is said to be "playing a chess game out there" or when routine political maneuvering is called "three-dimensional chess."

Pop culture has long been obsessed with chess as an indicator of brilliance and strategy. Hollywood tough guys Humphrey Bogart and John Wayne were both chess aficionados and played on the set with and without the cameras rolling. My favorite James Bond film, *From Russia with Love*, contains no small amount of chess. Early on, one of Bond's associates warns him, "These Russians are great chess players. When they wish to execute a plot, they execute it brilliantly. The game is planned minutely; the gambits of the enemy are provided for."

The end of the Cold War and the passing of the era of Russians as the bad guys in every movie didn't put an end to pop culture's affinity for an ancient board game. Many of today's top franchises highlight chess scenes. The *X-Men* movies put Professor X and Magneto across a glass board and set. *Harry Potter* has its Wizard's Chess, whose animated pieces are reminiscent of the game between C-3PO and Chewbacca in *Star Wars*. Even heartthrob vampires play chess, as seen in the Twilight movie *Breaking Dawn*.

Chess-playing machines have also figured prominently in fiction. In Stanley Kubrick's 1968 film, *2001*, the computer HAL 9000 easily defeats the character Frank Poole, foreshadowing that the machine will eventually murder him. Kubrick loved chess, so the game in his movie, like the one at the start of *From Russia with Love*, was based on a historical tournament game. Arthur C. Clarke's *2001* novel doesn't include a game, but it does mention that HAL could easily beat any of the humans on the ship if it played at full strength, but since that would be bad for morale it had been programmed to only win 50 percent of the time. Clarke adds, "His human partners pretended not to know this."

Advertisers are paid to exploit the power of symbols and again we see chess routinely deployed as a winning metaphor. Chess imagery in ads for banks, consultancies, and insurance companies seems obvious enough, but what about in commercials for Honda trucks, billboards for BMW cars, and online ads for dating websites? When you consider that only an estimated 15 percent of the US population plays chess, its cultural prominence is extraordinary.

It is also paradoxically at odds with the negative stereotypes of chess players as socially stunted, as if our brains developed processing power at the expense of emotional intelligence. It is true that chess can be a refuge for quiet people who prefer the company of their own thoughts, and obviously it doesn't require teamwork or social skills to excel. And even in the tech-obsessed twenty-first century, where Silicon Valley is Shangri-la and where it has become conventional wisdom that the geeks and nerds are the big winners, a particularly American strain of anti-intellectualism still bubbles up regularly.

Much of this fetishizing of chess and its practitioners, pro and con, stems from a simple lack of familiarity with the game. Relatively few Westerners play chess at all and fewer play to a level beyond knowing the rules. I've noticed that games without a chance factor—rolled dice, shuffled cards—are often considered hard, more like work than relaxing fun. Along with having no luck element, chess is a 100 percent information game; both sides know everything about the position all the time. There are no excuses in chess, no guesses, nothing out of the players' control.

Because of these factors, chess mercilessly punishes disparities in skill level, making it less friendly to newcomers who often don't have opponents of similar level at hand. After all, nobody likes to lose every time, as HAL's programmers realized. Poker and backgammon are games of skill, but their luck element is strong enough for every player to credibly dream about an upset in any given match. Not so with chess.

Chess-playing software on PCs and mobile devices and the Internet has mitigated this problem by providing a ready supply of opponents of all levels with 24/7 availability, although this also puts chess into direct competition with the never-ending supply of new online games and diversions. It also poses an interesting chess Turing test since you have no way to be sure whether you are playing against a computer or a human when you play online. Most people are far more engaged when playing against other humans and find facing computer opponents a sterile experience even when the machine has been dumbed down to a competitive level.

While chess programs today are so strong it's hard to tell the difference between their games and those of elite human Grandmasters, it has proved difficult to create convincingly weak chess machines. They tend to alternate between strong play and grotesque blunders during the same game. It's more than a little ironic that after half a century of trying to build the strongest chess entity on Earth, the programmers today are more concerned about making them play worse. Unfortunately, Arthur C. Clarke did not provide any guidance on how HAL arrived at its programmed mediocrity.

As a side note, it's a little curious that we take such joy and pride in winning a game due to a lucky roll or hand, is it not? I suppose it is human nature to revel in good fortune and beating the odds, merited or not, and everyone loves an underdog. Still, the phrase "it's better to be lucky than good" must be one of the most ridiculous homilies ever uttered. In nearly any competitive endeavor, you have to be damned good before luck can be of any use to you at all.

I WAS VERY INTERESTED in improving chess's image in the West even before I became world champion in 1985, and I did my best to speak out against the negative stereotypes of chess and chess players. I was also aware of the power of my own example in this regard, and in interviews and press conferences made a conscious effort to present myself as a well-rounded human being with interests beyond the sixty-four squares. This wasn't hard, since I was very much interested in history and politics, among other things, but as often as not, the articles about me in the mainstream press still fixated on angles that made me and other Grandmasters sound abnormal instead of like normal people with a particular talent.

There are practical and social considerations at work, as with every stereotype, and cultural traditions change very slowly. For better or worse, chess has been broadly categorized in the West as a slow and difficult game, reserved for smart people and bookworms at best, for misanthropic nerds at worst. This image is being refuted at the grassroots level thanks to the rising popularity of scholastic chess programs. After all, how can a game easily learned and greatly enjoyed by six-year-olds be difficult or dull?

In the Soviet Union, where I was raised and where chess was officially promoted as a national pastime, chess possessed less mystique and was treated as a professional sport. Soviet chess masters and instructors were accorded respect and a decent living. Nearly every citizen learned to play, and having such a large base of players meant finding more top talents, who were given special training. The game had deep Russian roots during tsarist times and, after the 1917 revolution, was prioritized by the Bolsheviks with the goal of endowing the new proletariat society with intellectual and martial values. As early as 1920, special military exemptions were given to strong chess players so that they could play in Moscow in the first Soviet Russia championship instead of being sent to the civil war front.

Years later, Joseph Stalin, though not much of a chess player himself, continued to support and promote the game as a way of demonstrating to the world the superiority of the Soviet man and the Communist system that produced him. While I cannot agree with that conclusion, you cannot argue with the results chesswise, as the Soviet Union completely dominated world chess for decades, winning the gold medal in eighteen of the nineteen Chess Olympiads in which it participated from 1952 to 1990. The world championship was held by five different Soviets starting with the first post-WWII championship contest in 1948 until 1972, and then again from 1975 until the impending collapse of the USSR, which allowed me to proudly exchange my Soviet flag for a Russian one hastily handmade by my mother, Klara, for my 1990 world championship match with Anatoly Karpov in New York City.

My own coming of age as a serious chess player in Baku, Azerbaijan, was benefited by this renaissance of political interest in chess in the 1970s. The Soviet leadership had been put into a panic by the avalanche of victories by American Bobby Fischer over the leading Soviet players. When Fischer took the world championship title from Boris Spassky in 1972 it became a matter of national pride to find and train players who could retake the crown. This happened sooner than expected when Fischer declined to defend his title in 1975 and it was given to Karpov by forfeit.

I was recruited into the Soviet chess machine at a very young age and given coaching and a place in the school of former world champion Mikhail Botvinnik. The "Patriarch of the Soviet Chess School," as Botvinnik was rightly called, also figures into the history of computer chess. An engineer by training, Botvinnik spent much of his retirement from chess working with a group of Soviet programmers to develop a chess program, an endeavor that resulted in nearly complete failure.

And so to me, playing chess was a completely normal thing to do both as a career and as recreation. As a young star I was allowed to travel abroad for tournaments and there I encountered for the first time the strange prejudices about chess players as eccentric geniuses or mentally unstable savants. It made no sense to me at all. I knew dozens of elite players and they were, if not "normal," whatever that means, all quite different from one another. Even selecting only from the world champions, they ranged from the mellow musicality of Vasily Smyslov to the chain-smoking and wisecracking of Mikhail Tal. Botvinnik was a stern professional from dawn to dusk in his suit and tie while Spassky had the air of a bon vivant and would occasionally show up to his games in tennis whites.

My own nemesis for five consecutive world championship matches, Karpov, was considered ice to my fire, both on and off the board. His soft-spoken demeanor and dependable character matched his quiet, boa constrictor chess style, while my exuberance and outspokenness mirrored my dynamic attacking play. The only thing all of us had in common was being very good at chess.

AS OFTEN HAPPENS, a few prominent cases from fiction and from real life helped create a lasting stereotype. The American chess champion Paul Morphy of New Orleans was also likely the first American world champion in any discipline after crushing Europe's best players on a tour in 1857–58. Soon after his hero's welcome he left chess to make his way as a lawyer, only to struggle and later suffer mental breakdowns that many attributed, without evidence, to the strain of his chess exploits.

The next American world champion, Bobby Fischer, is more recent and his decline is better documented. Fischer wrested the world championship title away from Boris Spassky and the Soviet Union in a legendary match held in Reykjavik, Iceland, in 1972. Partially due to Fischer's outrageous behavior leading up to and during the "match of the century," the international media coverage was incredible. Each game of the Cold War showdown was shown live around the world, even on American television. I was nine years old and already a strong club player when the Fischer-Spassky match took place and I avidly followed the games. Fischer, who had crushed two other Soviet Grandmasters on his march to the title match, nonetheless had many fans in the USSR. They respected his chess, of course, but many of us quietly enjoyed his individuality and independence.

After the match ended in a convincing victory for the American the world was at his feet. Chess was on the cusp of becoming a commercially successful sport for the first time. Fischer's play, nationality, and charisma created a unique opportunity. He was a national hero whose popularity rivaled that of Muhammad Ali. (Would the secretary of state have called Ali before a fight the way Henry Kissinger called Fischer in 1972?)

With glory comes responsibility and tremendous pressure. Fischer couldn't bring himself to play again. He spent three years away from the board before the precious title he had worked his entire life for was forfeited without the push of a pawn in 1975. Astronomical amounts of money were offered to bring him back. He could have played a match against the new champion, Karpov, for an unheard of $5 million. Opportunities abounded, but Fischer's was a purely destructive force. He demolished the Soviet chess machine, but could build nothing in its place. He was the ideal challenger and a disastrous champion.

When Fischer was lured out to play a so-called championship rematch with Spassky in Yugoslavia, then under UN sanctions, in 1992, his predictably rusty chess was accompanied by vociferous anti-Semitic and anti-American paranoia. He surfaced infrequently after that, each time causing the chess world to cringe and brace itself. Fischer's recorded rants rejoicing over the terror attacks on 9/11 could have done serious damage to the image of chess and chess players had they been more widely heard.

Fischer died alone in Iceland in 2008, having been offered refuge by the host of his greatest triumph. I am still asked about him regularly and no, I never played him or even met him. Everyone is keen to diagnose everything from schizophrenia to Asperger's from afar, a foolish and dangerous practice to be sure. I will say only that I am certain it was not chess that drove Fischer mad, if indeed he ever was mad. Fischer's tragic downfall wasn't what happens when someone plays chess; it's what happens when a fragile mind leaves his life's work behind.

I CANNOT DENY that the many legends and metaphors around the game have benefited me and my reputation. As much as I like to be appreciated for my work in human rights, my lectures and seminars to business and academic audiences, my foundation's work in education, and my books on decision making and Russia, I recognize that "former world chess champion" is a calling card with few peers. And, as I explained in detail in that 2007 book on decision making, *How Life Imitates Chess*, my chess career shaped and informed my thinking in every way.

I was just twenty-two years old when I became world champion in 1985, the youngest champion ever. My precocity created an awkward dynamic for me and my interviewers, since few young stars in any discipline are aware of why they excel. Instead of talking mostly to the chess press about openings and endgames, suddenly I was receiving earnest questions about everything from Soviet politics to my diet and my sleep habits from *TIME, Der Spiegel,* and even *Playboy.* As hard as I tried, I'm sure my banal answers often disappointed them. There was no secret, only innate gifts, hard work, and discipline that I learned from my mother and Botvinnik.

During my professional career, there were a few moments when I had the chance to step back and consider where chess fit in the greater arc of my life and, perhaps, in the world, but I rarely had the opportunity to dig into these matters for long. It wasn't until I retired from professional chess in 2005 that I had time to think more deeply about thinking and to see chess as a lens through which to investigate the decision-making processes that define every second of our waking lives.

The exceptions that occurred during my chess career are very much at the root of this book. My matches against computers, which spanned nearly the entire twenty years I spent as the world's top-rated player, allowed me to think about chess as something other than a competition. Battling each new generation of chess machines meant participating in a hallowed scientific quest, sitting at the nexus of human and machine cognition, and holding up the banner for mankind.

I could have spurned these invitations, as many of my Grandmaster colleagues did, but I was fascinated by the challenge and by the experiment itself. What could we learn from a strong chess machine? If a computer could play world-championship-level chess, what else could it do? Were they intelligent and what did that really mean? Could machines think, and what did the answers tell us about our own minds? Some of these questions have been answered while others are more passionately disputed than ever.

CHAPTER 2

RISE OF THE CHESS MACHINES

IN 1968, when the *2001* book and movie were created, it was not yet a foregone conclusion that computers would come to dominate humans at chess, or anything else beyond rote automation and calculation. As you might expect from the dawn of the computer age, predictions about machine potential were all over the map. Utopian dreams about the fully automated world just around the corner shared column space with dystopian nightmares of, well, pretty much the same thing.

This is a critical point to keep in mind before we criticize or praise anyone for their predictions, and before we make our own. Every disruptive new technology, any resulting change in the dynamics of society, will produce a range of positive and negative effects and side effects that shift over time, often suddenly. Consider the most discussed impact of the machine age, employment. The avalanche of factory automation, business machines, and domestic labor-saving devices that, starting in the 1950s, led to the disappearance of millions of jobs and entire professions, while skyrocketing productivity created unprecedented economic growth—and the creation of more jobs than had been lost.

Should we pity all the steel-driving John Henrys put out of work by steam engines? Or the office pool typists, assembly-line workers, and elevator operators who had to retool and retrain as technology replaced them by the thousands? Or should we consider them lucky for being able to leave behind such work, work that is tedious, or physically exhausting, or dangerous?

Our attitude matters, and not because we can stop the march of technological progress even if we wanted to, but because our perspective on disruption affects how well prepared for it we will be. There is plenty of room between the utopian and dystopian visions of the fully automated and artificially intelligent future we are heading into at rapidly increasing speed. Each of us has a choice to make: to embrace these new challenges, or to resist them. Will we help shape the future and set the terms of our relationship with new technology or will we let others force the terms on us?

JUST AS I was fascinated by chess machines, generations of scientific luminaries have been fascinated with chess and with making machines that played chess. You might assume that the mathematicians, physicists, and engineers who formed the first wave of computer scientists and cyberneticists in the 1950s would hold little romanticism for a board game, even one they enjoyed passionately. And yet several of these eminently logical, scientific minds insisted that if a machine could be taught to play chess well, surely the secrets of human cognition would be unlocked at last.

This sort of thinking is a trap into which every generation falls when it comes to machine intelligence. We confuse performance—the ability of a machine to replicate or surpass the results of a human—with method, how those results are achieved. This fallacy has proved irresistible in the domain of higher intelligence that is unique to *Homo sapiens*.

There are actually two separate but related versions of the fallacy. The first is "the only way a machine will ever be able to do X is if it reaches a level of general intelligence close to a human's." The second, "if we can make a machine that can do X as well as a human, we will have figured out something very profound about the nature of intelligence."

This romanticizing and anthropomorphizing of machine intelligence is natural. It's logical to look at available models when building something, and what better model for intelligence than the human mind? But time and again, attempts to make machines that think like humans have failed, while machines that prioritize results over method have succeeded.

Machines don't need to do things the same way the natural world does in order to be useful, or to surpass nature. This is obvious from millennia of physical technology and it applies to software and artificially intelligent machines as well. Airplanes don't flap their wings and helicopters don't need wings at all. The wheel doesn't exist in nature, but it has served us very well. So why should computer brains work like human brains in order to achieve results? As is so often the case in the crossroads of human and machine thinking, chess proved to be an ideal laboratory for investigating this question.

Beyond science fiction, the matter of whether a machine can be intelligent didn't really arise among technologists and the general public until the digital took over from the mechanical and analog in the 1940s and vacuum tubes gave way to semiconductors in the 1950s. It was as if ghosts could be imagined in the machines as soon as their processes could no longer be followed by the naked eye. Mechanical calculators had been around since the seventeenth century and key-driven desktop versions were produced in the thousands by the middle of the nineteenth. Programmable mechanical calculators were designed by Charles Babbage in 1834, and the first "computer" program for one was written by Ada Lovelace in 1843.

Despite the impressive sophistication of these machines, nobody seriously wondered if they were intelligent any more than they did about pocket watches or steam locomotives. Even if you had no idea how a mechanical device like a cash register performed, you could hear the wheels spinning. You could open it up and see the gears turning. As amazing it was for a machine to perform "mental" feats like logic and mathematics faster than a human could, there was little discussion of how it did it compared to how the human mind worked.

This was due partly to the relatively comprehensible nature of these early machines and partly because human cognition wasn't very well understood. We'd come a long way since the fourth century BC, when Aristotle believed the brain was a sort of cooling organ while the senses and intelligence resided in the heart, something to remember the next time you hear the phrase "learn something by heart." But it wasn't until toward the end of the nineteenth century, with the discovery of neurons, that the idea of the brain as an electrically powered calculation device became possible. Before that, the concept of the brain was more metaphysical than physical, with Roman-era arguments about "animal spirits" and where, exactly, the soul resided.

Souls aside, it is generally agreed today that the mind is not greater than the sum of a being's physical parts and experiences. The mind goes beyond reasoning to include perception, feeling, remembering, and, perhaps most distinctively, *willing*—having and expressing wishes and desires. Brains grown in petri dishes from stem cells are interesting for experiments, but without any input or output they could never be called minds.

WHEN YOU LOOK BACK at the history of computers it seems like as soon as a machine is invented, the next step is to turn it into a chess player. For the first decades of computing, chess was always near the forefront. Along with the reputation of the game, many of the founding fathers of computation were dedicated chess players, so they were quick to see the game's potential as a challenging test bed for their programming theories and electronic inventions.

How do machines play chess? The basic formula hasn't changed since 1949, when the American mathematician and engineer Claude Shannon wrote a paper describing how it might be done. In "Programming a Computer for Playing Chess," he proposed a "computing routine or 'program'" for use on the sort of general-purpose computer Alan Turing had theorized years earlier. You can tell how early it was in the computer age that Shannon put the word "program" in quotation marks as jargon.

As with many who followed him, Shannon was slightly apologetic at proposing a chess-playing device of "perhaps no practical importance." But he saw the theoretical value of such a machine in other areas, from routing phone calls to language translation. Shannon also explained as well as anyone why chess was such an excellent test bed:

The chess machine is an ideal one to start with, since

- the problem is sharply defined both in allowed operations (the moves) and in the ultimate goal (checkmate);
- it is neither so simple as to be trivial nor too difficult for satisfactory solution;
- chess is generally considered to require "thinking" for skillful play; a solution of this problem will force us either to admit the possibility of a mechanized thinking or to further restrict our concept of "thinking";
- the discrete structure of chess fits well into the digital nature of modern computers.

Pay particular attention to point three, where Shannon bridges the gap between computer science and the metaphysical world in just thirty-five words. Since chess requires thinking, either a chess-playing machine thinks or thinking doesn't mean what we believe it to mean. I also admire his use of the word "skillful," since simply memorizing the rules and making random legal moves or regurgitating moves from memory (or a database) isn't how he defines thinking.

This insight echoes Norbert Wiener's note at the end of his seminal 1948 book, *Cybernetics*: "Whether it is possible to construct a chess-playing machine, and whether this sort of ability represents an essential difference between the potentialities of the machine and the mind."

Shannon went on to describe the various factors a chess program would need, including the rules, piece values, an evaluation function, and, most critically, the possible search methods a future chess machine could use. He described the most fundamental element of search, what we call the "minimax" algorithm, which originated in game theory and has been applied to logical decision making in many fields. Very simply put, a minimax system evaluates possibilities and sorts them from best to worst.

In games like chess, the program uses its evaluation system to rate as many variations as possible in the given position and puts a value on each position it sees. The move that returns the highest value is put at the top of its move list as the move to make. The program has to evaluate all the possible moves of both players, as deeply as time allows.

In an important contribution, Shannon outlined "Type A" and "Type B" search techniques. This is rather boring nomenclature, to be honest, and it's probably helpful to think of Type A as "brute force" and Type B as "intelligent search." Type A is an exhaustive search method that examines every possible move and variation, deeper and deeper with each pass. Type B describes a relatively efficient algorithm that works more like the way a human player thinks by focusing only on a few good moves and looking deeply at those instead of checking everything.

Think about selecting a chess move the way you choose a pastry at a bakery with a long glass case. You don't need to look at every single item in the case before you order, and even if you do, you don't need to ask what every item is and what its ingredients are. You know what type of pastries you like best, what they look like and taste like. You quickly narrow your choice down to a few favorites before taking time to decide among them.

But wait! You spy something in the corner of the case you haven't seen before and it looks quite delicious. Now you have to slow down a little, maybe ask the clerk for more information about it, and use your evaluation function to find out if it's something you'd actually enjoy. Why did it look delicious? Because it's in some way analogous to something you have had before and liked. This is also how strong human chess players start evaluating moves even before we start doing any calculation. The pattern-matching part of the brain has rung a bell to attract our attention to something interesting.

At the risk of overextending this analogy and also making you hungry, the bakery itself matters as well. If it's the same bakery you go to every day, your choice is nearly automatic, perhaps based on the time of day or what you're in the mood for. But what if it's a bakery you've never been to before, in a country you're visiting for the first time? You don't recognize anything; your intuition and experience are practically worthless. Now you have to use brute force, a Type A search, asking about each item, each ingredient, and trying samples before you decide. You may still get something you like, but it takes much more time to make a quality decision this way.

That describes a novice human chess player and, to a degree, a stronger one in a chaotic and completely new position. But chess is a limited game and every position will have patterns and markers our intuition can interpret. Each of the estimated tens of thousands of positions a strong master has imprinted in memory can also be broken down into component parts, rotated, twisted, and still be useful. Outside of the opening sequences that are indeed memorized, strong human players don't rely on recall as much as on a super-fast analogy engine.

When I look at a chess position, whether it's my own game or someone else's, there is very little that is consciously systematic about my move search process. Some moves are forced, meaning either legally obligated, as in the case of a check when your king is attacked, or when every other move clearly loses. This happens regularly throughout the game, such as when a piece is captured and you must recapture or face a big material deficit. Some games contain several dozen forced moves, and almost no real search is needed on those moves. Just like you don't have to tell yourself consciously not to walk into traffic, these moves are practically reflex for a competent player.

Disregarding forced moves, each position will have three or four plausible moves, sometimes as many as ten or so. Again, before any real search begins in my mind, I have selected several to analyze more deeply, what we call candidate moves. Of course, I'm not starting from scratch if it's my own game; I've been planning my strategy and looking at the most likely variations during my opponent's time on the clock. If he makes the move I was expecting it's quite possible I will reply instantly. And often I will plan out a sequence of four or five moves in advance, only pausing to double-check my calculations if the sequence plays out as expected.

Most of my search and evaluation time is spent on the main variation, the move I selected as the most likely right at the start. My calculation skills are attempting to validate my intuition. If my opponent's move was a surprise, something I never considered during my time pondering on his move, I may take some extra time to peruse the whole board for new weaknesses and opportunities.

The human mind isn't a computer; it cannot progress in an orderly fashion down a list of candidate moves and rank them by a score down to the hundredth of a pawn the way a chess machine does. Even the most disciplined human mind wanders in the heat of competition. This is both a weakness and a strength of human cognition. Sometimes these undisciplined wanderings only weaken your analysis. Other times they lead to inspiration, to beautiful or paradoxical moves that were not on your initial list of candidates.

I wrote about how intuitive flights of fantasy can cut through the fog of calculation in *How Life Imitates Chess*, and I cannot resist sharing here the inimitable storytelling of eighth world champion Mikhail Tal, known as the "Magician from Riga" for his dazzling tactical imagination at the board. In this self-interview in his 1976 book, Tal is discussing what was going through his head while he was contemplating a knight sacrifice in a game against another Soviet Grandmaster.

> Ideas piled up one after another. I would transport a subtle reply to my opponent, which worked in one case, to another situation where it would naturally prove quite useless. As a result, my head became filled with a completely chaotic pile of all sorts of moves, and the famous "tree of variations," from which the trainers recommend that you cut off the small branches, in this case spread with unbelievable rapidity.
>
> And then suddenly, for some reason, I remembered the classic couplet by [well-known Soviet children's poet] Korney Chukovsky:
>
> *Oh, what a difficult job it was*
> *To drag out of the marsh the hippopotamus.*
>
> I don't know from what associations the hippopotamus got onto the chess board, but although the spectators were convinced that I was continuing to study the position, I was trying at this time to work out: Just how would you drag a hippopotamus out of the marsh? I remember how

> jacks figured in my thoughts, as well as levers, helicopters, even a rope ladder. After a lengthy consideration, I admitted defeat as an engineer, and thought spitefully, "Well, let it drown!"
>
> And suddenly the hippopotamus disappeared. Went off from the chess board just as he had come on. Of his own accord. And straightaway the position did not appear to be so complicated. Now I somehow realized that it was not possible to calculate all the variations, and that the knight sacrifice was, by its very nature, purely intuitive. And since it promised an interesting game, I could not refrain from making it.
>
> And the following day, it was with pleasure that I read in the paper how Mikhail Tal, after carefully thinking over the position for 40 minutes, made an accurately calculated piece sacrifice.

Tal was a man of rare humor and honesty as well as chess brilliancy. Concentration and mental organization are essential for professional chess players, but I suspect that we rely on such intuitive leaps more often than we would like to admit.

A game of chess is an intense competition, not a laboratory experiment. Under pressure, with a ticking clock, mental discipline breaks down. Visualization becomes imperfect, even for Grandmasters, and blunders become more likely. Sometimes you spend ten minutes on your main variation only to find out that it is a fatal mistake. Panic! Despair! Or after your opponent's move you see what looks at first like a brilliant winning coup. Elation! But do you have another ten minutes to invest in order to confirm your instincts? Do you just play it anyway, hoping your intuition hasn't led you astray? Of course, computers don't have to worry about any of these psychological dramas, which is as much a reason they are so tough to play against as how many millions of positions they analyze per second.

Returning to 1949, Claude Shannon held out little hope for the success of Type A programs that would have to analyze every possible move in deeper and deeper iterations. The numbers just didn't seem feasible. He lamented that even if a Type A machine evaluated one position per microsecond ("very optimistic"), it would take more than sixteen minutes per move, or ten hours for its half of a typical forty-move game. And it would still be very weak because that would only allow it to see three moves deep in its exhaustive search tree, only enough to beat a very weak human player.

The main problem of chess programming is the very large number of possible continuations involved, what is called the "branching factor." Right from the start, the sheer number of possibilities was enough to stress the resources of the fastest computers then conceivable. Each side starts with a force of sixteen, eight pieces and eight pawns. There are over 300 billion possible ways to play just the first four moves in a game of chess, and even if 95 percent of these variations are terrible, a Type A program would still have to check them all in order to be sure.

It gets worse. In an average position there are around forty legal moves. So if you consider every reply to each move, you already have sixteen hundred moves to evaluate. This is after just two "ply," as programmers call half-moves, one by white and one by black. After two moves each (four ply) there are 2.5 million; after three moves it's 4.1 billion. The average game lasts forty moves, leading to numbers that are beyond astronomical. The total number of legal positions in a game of chess is comparable to the number of atoms in our solar system.

And so Shannon, a decent and well-read player himself, put his hopes in a Type B strategy that would think more selectively and so more efficiently. Instead of looking at every possible position and every variation to equal depth, a Type B algorithm would operate the way a good human player does by concentrating on the most plausible and most forcing moves and then working those out deeply while discarding the implausible moves at the start.

Human players learn very quickly that only a handful of moves make sense, and the stronger the player the faster and more accurately this initial sorting and sifting is done. Beginners are more like Type A computers in that they tend to look all over the board comprehensively, relying on brute force to calculate the consequences of each move. This method works for a computer that looks at millions of positions per second, but humans can't process like this. Even the human world champion can only see an estimated two or three positions per second.

If you manage to find the four to five most reasonable moves in a given position and discard the rest, which is not trivial at all, the geometric branching of the decision tree still becomes enormous very quickly. So even if you succeed in creating a Type B algorithm that can search more intelligently, you still need a lot of processing speed and a lot of memory to keep track of all those millions of position evaluations.

I've already mentioned Alan Turing's "paper machine," the first known functional chess program. I even had the honor of playing a reconstructed version of it on a modern computer when I was invited to speak at the Turing centenary in Manchester in 2012. It was quite weak by modern standards, but still must be considered a remarkable achievement considering that Turing didn't even have a computer to test it on.

When computers capable of running chess code finally came along a few years later, they were so dismally slow that it was assumed that Shannon was right, and that the best hope for real progress was Type B. It was a logical conclusion, since machines that could search Shannon's optimistic one-position-per-microsecond benchmark were still decades away. Any program that looked at every possible move would take weeks to reach the search depth required to play a rational game and years to play well. But as it turned out, and not for the last time, the assumption that humanlike was better than brute force was largely wrong.

IN 1956, the nuclear laboratory of Los Alamos was the site of the next advance in chess computing, taking the theories of Wiener, Turing, and Shannon and turning them into an actual chess-playing machine. One of the first computers, the gigantic MANIAC 1, had twenty-four hundred vacuum tubes and the revolutionary ability to store programs in memory. As soon as it was delivered, some of the H-bomb scientists tested it out by writing a chess program. Of course! The computer's resources were so limited that they had to use a reduced board, just six-by-six squares, without bishops. After playing against itself and then losing to a strong player (despite the human playing without a queen), the machine beat a young volunteer who had just learned the game. It didn't make headlines, but it was the first time a human had lost to a computer in a game of intellectual skill.

Just one year on from that landmark, in 1957, a group of researchers at Carnegie Mellon University proclaimed that they had discovered the secret to a Type B–style chess algorithm that would lead to a machine defeating the world champion in just ten years' time. Considering how slow computers were then, and how expensive, this was nearly as bold as John F. Kennedy's declaration in 1962 that the United States would put a man on the moon by the end of the decade.

Or perhaps it was simply uninformed and wildly unrealistic. Even had the entire industrial might of America been put into beating the world chess champion by 1967, their prediction almost certainly would not have come to pass. The Apollo program required the creation of new materials and novel technologies, and JFK's goal was achieved only by pushing the limits of nearly every constituent technology. Still, it was an achievement of its time, conceived and developed on a relatively predictable timeline. Those in charge of the Apollo program in 1962 understood what they would have to do to put humans on the moon, if not exactly how to do it.

In contrast, a world-champion-caliber chess machine didn't exist until 1997, thirty years after the Carnegie Mellon team's predicted date, despite computer power doubling every two years, roughly in accord with Moore's law. It was soon clear that their killer "smart" algorithm was fatally flawed and that they weren't really sure what the best path forward would be. Chess was too complex; computers were too slow. A few million more person-hours dedicated to chess algorithms in the 1960s surely would have made great advances in programming knowledge and hardware design, but the computer hardware necessary to store and run such sophisticated programs at speeds fast enough to beat a Grandmaster wouldn't exist until the 1980s.

Even had the equivalent of NASA's budget been invested at the time, a world-beating program by 1967 would have been unimaginable and even by 1977 is quite dubious. The Cray-1 supercomputer installed at Los Alamos National Laboratory in 1976 was the fastest computer in the world with a speed of 160 million operations per second (160 megaflops). In comparison, the Deep Junior program I played to a draw in a match in 2003 ran on four Pentium 4 chips that were each roughly twenty times faster than the Cray-1, and it already played as well or better than Deep Blue did in 1997 on its specialized hardware.

This wasn't because Deep Junior was faster than Deep Blue; it wasn't. In fact, Deep Blue looked at an average of fifty times as many positions per second, 150 million to 3 million. But raw speed is only one factor in a machine's chess strength. The efficiency of the programming is critical for getting the most out of the hardware. Designing smarter search routines and making steady optimizations in the code are where most of the gains in a program's chess strength come from, according to several generations of chess programmers going back to the 1970s.

The trade-offs come when the programmer has to add chess knowledge to the machine's search algorithm. The most basic chess program must understand the concept of checkmate, for example, and the relative values of the pieces. If you tell the machine that rooks and bishops are both worth three pawns, when in fact rooks are more powerful than bishops, it's not going to play very well. Counting material, who has more pieces and pawns, is something chess machines do very quickly and very well. And it doesn't take a lot of chess knowledge on the part of the programmer to assign these standard values.

After the material value of the pieces and pawns, you have more abstract knowledge such as which player controls more space on the board, the structure of the pawns, and king safety. Every time you give the computer another piece of information to evaluate on every move, the search becomes slower. In sum, a chess program can either be faster and dumber or slower and smarter. It's a fascinating balancing act, and it took decades to create machines that were both smart enough and fast enough to challenge the world's best human players.

HOWEVER POOR the early predictions were, there was steady progress over the next twenty years. Trial and error in programming techniques and the relentlessness of Moore's law produced chess machines that played at the level of the top 5 percent of human players by 1977, expert level. They still played terrible chess, full of illogical moves even a weak human would never consider. But they were becoming fast enough to cover up these occasional blunders with accurate defense and sharp tactics while playing against humans.

Faster hardware was only one part of their progress. Most of the rest came from better programming, speeding up the search algorithm. The "alpha-beta" algorithm allowed the programs to rapidly prune out weak moves and thus see further ahead, faster. This was an evolution of the minimax algorithm described by Shannon as Type A, or brute force. The program stops focusing on any move that returns a lower value than the currently selected move. With this key improvement and other optimizations, Type A programs became ascendant over Type B. Efficient brute force was dominant over every attempt to emulate human-style thinking and intuition in chess machines. Some chess knowledge was still necessary, but speed was king.

All modern chess programs are based on applying this alpha-beta pruning search algorithm to the basic minimax concept. On this structure, the programmers build the chess evaluation function, tuning it for optimal results. The first programs using this technique, running on some of the fastest computers of the day, reached a respectable playing strength. By the late 1970s, programs running on early personal computers like the TRS-80 could defeat most amateurs.

The next leap came out of the famous Bell Laboratories in New Jersey, which churned out patents and Nobel Prize winners for decades. Ken Thompson built a special-purpose chess machine with hundreds of chips. His machine, Belle, was able to search about 180,000 positions per second while the general-purpose supercomputers of the day could only manage 5,000. Seeing up to nine half-moves (ply) ahead during a game, Belle could play at the level of a human master and far better than any other chess machine. It won just about every computer chess event from 1980 to 1983, before it was finally surpassed by a program running on the next generation of Cray supercomputers.

Consumer chess programs with names like Sargon and Chessmaster continued to improve while benefiting from the rapid increase in processor speeds provided by Intel and AMD. Then specialized hardware in the mold of Belle made a comeback thanks to a new generation of chess machines designed at Carnegie Mellon. Professor Hans Berliner was a computer scientist as well as a world champion at correspondence chess (chess played by mail, now typically email). His team's machine HiTech hit a milestone by reaching a Grandmaster rating in 1988, but it was soon bettered by the creation of his graduate students Murray Campbell and Feng-hsiung Hsu. Their specialized hardware machine Deep Thought became the first chess machine to defeat a Grandmaster in a regular tournament game in November 1988. Upon graduating in 1989, they took Deep Thought and joined IBM, where their project was rechristened to reflect the company's "Big Blue" nickname. Deep Thought became Deep Blue and the last great chapter of the machine chess story began.

CHAPTER 3

HUMAN VERSUS MACHINE

HUMAN COMPETITION with machines has been part of the conversation about technology since the first machines were invented. We continue to update the terminology, but the basic narrative remains the same. People are being replaced, or are losing a race, or being made redundant, because technology is doing what humans used to do. This "human versus machine" narrative framework arose to prominence during the industrial revolution, when the steam engine and mechanized automation in agriculture and manufacturing began to appear on a large scale.

The competition story line grew more ominous and more pervasive during the robotics revolution of the 1960s and 1970s, when more precise and more intelligent machines began to encroach on the jobs of people with more powerful social and political representation, like unions. The information revolution came next, culling millions of jobs from the service and support industries.

Now we have reached the next chapter in the human versus machine employment story, when the machines "threaten" the class of people who write articles about it. We read headlines every day about how the machines are coming for the lawyers, bankers, doctors, and other white-collar professionals. And make no mistake, they are. Every profession will eventually feel this pressure, and it must, or else it will mean humanity has ceased to make progress. We can either see these changes as a robotic hand closing around our necks or one that can lift us up higher than we can reach on our own, as has always been the case.

Romanticizing the loss of jobs to technology is little better than complaining that antibiotics put too many grave diggers out of work. The transfer of labor from humans to our inventions is nothing less than the history of civilization. It is inseparable from centuries of rising living standards and improvements in human rights. What a luxury to sit in a climate-controlled room with access to the sum of human knowledge on a device in your pocket and lament how we don't work with our hands anymore! There are still plenty of places in the world where people work with their hands all day, and also live without clean water and modern medicine. They are literally dying from a lack of technology.

It's not just college-educated professionals who are under pressure today. Call center employees in India are losing their jobs to artificially intelligent agents. Electronics assembly-line workers in China are being replaced by robots at a rate that would shock even Detroit. There is an entire generation of workers in the developing world who were often the first in their families to escape farming and other subsistence labor. Will they have to return to the fields? Some may, but for the vast majority this isn't an option. It's like asking if all the lawyers and doctors will have to "return to the factories" that don't exist anymore. There is no back, only forward.

We don't get to pick and choose when technological progress stops, or where. Companies are globalized and labor is becoming nearly as fluid as capital. People whose jobs are on the chopping block of automation are afraid that the current wave of tech will impoverish them, but they also depend on the next wave of technology to generate the economic growth that is the only way to create sustainable new jobs. Even if it were possible to mandate slowing down the development and implementation of intelligent machines (how?), it would only ease the pain for a few for a little while and make the situations worse for everyone in the long run.

Unfortunately, there is a long tradition of politicians and CEOs sacrificing the long term and the greater good in order to satisfy a small constituency at the moment. Educating and retraining a workforce to adapt to change is far more effective than trying to preserve that workforce in some sort of Luddite bubble. But that takes planning and sacrifice, words more associated with a game of chess than with today's leaders.

Donald Trump won the US presidency in 2016 with promises of "bringing jobs back" from Mexico and China, as if American workers can or should be competing for manufacturing jobs with countries where salaries are a fraction of those in the United States. Putting high tariffs on foreign-made products would make nearly every consumer good far more expensive for those who can least afford such an impact. If Apple offered a red, white, and blue iPhone made in the United States that cost twice as much as the same model made in China, how many would they sell? You can't discard the downsides of globalization while keeping the benefits.

It's a privilege to be able to focus on the negative potential of world-changing breakthroughs like artificial intelligence. As real as these issues may be, we will not solve them unless we keep innovating even more ambitiously, creating solutions and new problems, and yet more solutions, as we always have. The United States needs to replace the jobs being lost to automation, but it needs new jobs to build the future instead of trying to bring back jobs from the past. It can be done and it has been done before. Here I'm not referring to the 30 percent of Americans who lived on farms in 1920, down below 2 percent nearly a century later, but to a much more recent retooling.

The launch of the tiny Sputnik device by Sergey Korolyov on October 7, 1957, turned the space race into a sprint that lasted for decades. President Eisenhower immediately ordered all American projects to move up their timetables, which likely helped contribute to the failed launch of the first American satellite, Vanguard, in December 1957. The media dubbed the failure, seen live on television, "Flopnik," and the embarrassment drove the administration to push even harder for results.

The phrase "Sputnik moment" subsequently entered the national lexicon to represent any foreign accomplishment that serves to remind America that it is not without rival. For example, the OPEC oil embargo of the 1970s was supposed to be a Sputnik moment that would goad the United States into developing renewable energy. Then came Japanese manufacturing technology in the 1980s, the expanded European Union in the 1990s, and the rise of Asia in the last decade.

A more recent Sputnikian wakeup call to rouse the American giant was supposed to be the 2010 revelation that kids in Shanghai scored far better on standardized math, science, and reading tests than their peers in other nations. An October 13, 2016, *Washington Post* headline warned that "China has now eclipsed us in AI research." Perhaps this fact is not unrelated to those 2010 test scores. Yet another Sputnik moment? As you can see, the track record of Americans picking up any of these gauntlets is quite poor, except, of course, the original.

Inevitably, all these repetitions have trivialized the impact of Sputnik, which combined many of the day's real and imagined fears into a twenty-three-inch-diameter metal sphere. American editorial pages of the day were filled with wonder and dread at this shocking combination of Communist ideology and unmatched technology. Sputnik stoked American fires in the most primeval ways: creating fear and anger, and denting America's national ego and pride.

The United States responded. In 1958, three years before President John F. Kennedy boldly promised to put a man on the moon by the end of the decade, then–Senator Kennedy supported legislation called the National Defense Education Act, which directly funded science education across the country. The future engineers, technicians, and scientists produced by the program would form the generation that designed and built much of the digital world we live in today.

It is still an open question whether a national revitalization effort can be summoned, like Aladdin's genie, on demand. It is depressing to consider the thought that war and fear are necessary requirements to inspire united action since we are obviously better off in a world with as little as possible of both. But existential threats do focus the mind wonderfully, as Samuel Johnson said about an impending hanging. Any transformative effort on a national scale requires the focused minds of politicians, business leaders, and a plurality of citizens to support it.

In the 1970s, superior Japanese cars were bought by American consumers in the millions. Chinese graduates are enthusiastically welcomed into every American university and firm. In today's globalized world, technological competition has given way to the sense that we all benefit from someone, somewhere, doing things right, or at least doing them better. While this is no doubt better than no one doing it right anywhere, we cannot abandon the quest for scientific excellence in the United States. America still possesses the unique potential to innovate on a scale that can push the entire world economy forward. A world in which America is content with mediocrity is, literally, a much poorer world.

When questioned by Congress about the Soviet success, President Eisenhower's special assistant for science and technology, Dr. James Killian, also president of MIT, gave a cultural answer to a technical question: "There is no doubt that the Soviets have generated a respect and enthusiasm for science and engineering that has operated to give them a large supply of trained professionals in these fields." He was quoted in the December 1957 issue of the *Bulletin of the Atomic Scientists,* whose editors were even more critical of the American mindset that had allowed the Soviets to pull ahead in space, as this editorial comment in the same article made clear: "We have catered to desires for undisturbed comfort rather than focusing on larger goals and developing our potentialities."

This was a polite and professorial way of saying that Americans had become lazy, short-sighted, and unwilling to take the risks required to stay on the cutting edge of technology. I'm worried that this is where the United States is finding itself once again. Silicon Valley is still the greatest hub of innovation in the world and America possesses more of the conditions necessary for success than anywhere else. But when is the last time you heard about a government regulation that promoted innovation instead of trying to limit it?

I'm a firm a believer in the power of free enterprise to move the world forward. All that Soviet respect for science was no match for the American innovation machine once unleashed. The problem comes when the government is inhibiting innovation with overregulation and short-sighted policy. Trade wars and restrictive immigration regulations will limit America's ability to attract the best and brightest minds, minds needed for this and every forthcoming Sputnik moment.

FIGHTING TO THWART the impact of machine intelligence is like lobbying against electricity or rockets. Our machines will continue to make us healthier and richer if we use them wisely. They will also make us smarter. It's an interesting diversion to consider what the first invention was that directly increased human knowledge and our understanding of the world. Starting in the thirteenth century, grinding glass led to glasses, and eventually to the telescope and the microscope, tools of human enhancement that dramatically improved our ability to control our environment via improved navigation and medical research. Perhaps only the compass is an earlier invention that provided us with information otherwise difficult or impossible to obtain. The abacus, from the third millennium BC, is as much a method as a machine, but it is likely the first device to augment human intelligence. The alphabet, paper, and the printing press didn't exactly create knowledge, but performed the essential corollary task of preserving and distributing it, much as the Internet does.

My own experiences battling computers across a game board are the exception that proves the rule. We aren't competing against our machines, no matter how many human jobs they can do. We are competing with ourselves to create new challenges and to extend our capabilities and to improve our lives. In turn, these challenges will require even more capable machines and people to build them and train them and maintain them—until we can make machines that do those things too, and the cycle continues. If we feel like we are being surpassed by our own technology it's because we aren't pushing ourselves hard enough, aren't being ambitious enough in our goals and dreams. Instead of worrying about what machines can do, we should worry more about what they still cannot do.

I will say again that I am not unsympathetic to those whose lives and livelihoods have been negatively impacted by disruptive new technology. Few people in the world know better than I do what it's like to have your life's work threatened by a machine. No one was sure what would happen if and when a chess machine beat the world champion. Would there still be professional chess tournaments? Would there be sponsorship and media coverage of my world championship matches if people thought the best chess player in the world was a machine? Would people still play chess at all?

The answer to all of these questions turned out to be yes, thankfully, but these doomsday scenarios were one reason some in the chess community criticized my eagerness to participate in human versus machine events at all. I suppose I could have delayed the inevitable a little by declining, and forcing the programmers to challenge other top players. If a machine had beaten Anand or Karpov, the next players on the rating list after me at the time of my rematch with Deep Blue in May 1997, the story would have been "Nice, but would it beat Kasparov?" But that would only last until I was no longer the world champion, which happened in 2000, or until I was no longer ranked number one and I retired from chess, which happened in 2005. I was never one to duck a challenge, and being remembered as the first world champion to lose a match to a computer cannot be worse than being remembered as the first world champion to run away from a computer.

And I didn't *want* to run away. I was thrilled by these new trials, by the scientific pursuit, by the new avenues to promote chess, and, frankly, by the attention and the money that sometimes came with it all. Why should someone else be the first, for better or for worse? Why should I exchange a unique and historic role as a participant to become just another spectator?

Nor did I believe the apocalyptic predictions about what might happen if I lost a match to a machine. I was always optimistic about the future of chess in the digital age, and not because of the trite and imprecise "people still run footraces even though cars are faster" justifications that many were making at the time. John Henry aside, automobiles didn't make walking obsolete or put pedestrians out of work. Many things on Earth are faster than Usain Bolt's top speed of thirty miles per hour, from coyotes (40 m.p.h.) to kangaroos (44 m.p.h.). So what?

Chess is a very different matter from physical sports, as strong chess machines can directly and indirectly influence human play. You can think of them as more analogous to steroids and other forms of doping in physical sports, as an external augmentation with the potential to boost performance or to damage the sport if abused. Chess is concrete; a move or strategy employed by a computer can be exactly duplicated by a human. What if machines showed us that some of the most popular chess openings were bad, and how to beat them? Would we human players become the automatons ourselves, regurgitating the moves and ideas shown us by our machines? Would the winner be the player with the strongest computer at home? Would there be an epidemic of computer-assisted cheating? These were realistic and serious questions, and they still are, but these are not the same as the dismal fantasies about computers solving chess for good or making human versus human play obsolete.

As with nearly every new technology, for every potential downside there were many upsides to the increasing strength and availability of strong chess machines. I admit, however, that I was late in recognizing this. The first few generations of PC chess software, powered by what are called "chess engines" in our vernacular, were too weak to be very useful to professional players. The most popular programs were directed toward casual consumers and focused more on pretty 3-D boards or animated pieces than on the strength of the engine. Even as they got much stronger and became dangerous opponents in the early 1990s, the chess they played was ugly and inhuman, not very useful for serious training.

Instead, my early interest was in developing computer tools to help with my preparation and that of other serious players. Instead of digging through dozens of reference books and my stacks of notebooks full of analysis, a database of thousands of games could be searched in a few seconds and could also be easily updated. In 1985, I started discussing the creation of such an app with the German tech writer Frederic Friedel, who was a serious aficionado of computer chess. He and a programmer acquaintance, Matthias Wüllenweber, founded ChessBase in Hamburg and released the ground-breaking program of the same name in January 1987. And with that, an ancient board game was pulled into the information age, at least if you had an Atari ST. The ability to collect, organize, analyze, compare, and review games with just a few clicks was, as I put it at the time in 1987, as revolutionary for the study of chess as the printing press.

As for chess engines, by the early 1990s I had lost a few blitz games to the top PC programs and it was clear they were only going to keep getting stronger. Before that, back when home computers were still uncommon in most of the world, machine capabilities were often wildly over- and underestimated. There had been a few early theories, optimistic ones from my perspective, that the exponential branching factor of chess analysis would create a barrier at some point, but programming techniques and ever-faster CPUs kept the machines' ratings rising steadily.

I gradually understood that the proliferation of strong programs could greatly democratize the sport worldwide. My success in chess was as much a case of geography being destiny as it was my having natural talent and a determined mother. In the Soviet Union, I had easy access to chess books, magazines, coaches, and a ready supply of strong opponents. Nowhere else in the world could offer these advantages, except perhaps the former Yugoslavia. Other national chess powers also counted on longstanding chess traditions that provided the resources necessary for talent to develop.

The existence of a Grandmaster-level chess program available on an inexpensive personal computer upended that hierarchy. While not as good as an experienced human coach, it was far better than nothing. Combined with the Internet's ability to bring the game to every corner of the world, a shift was under way. The key factor in producing elite chess talent is finding it early, and thanks to strong computers this is now very easy to do just about anywhere. It's no coincidence that the current list of elite chess players contains many representatives of countries with little or no old chess traditions. Computers tend to have this impact in many ways, reducing the influence of dogma. Chess in China and India has been boosted by government support and local stars, but the ability to train with Grandmaster machines helped make their rise into the elite ranks startlingly quick. Previously, it was necessary to import Soviet coaches and host expensive international tournaments or send local players abroad to find strong competition. China currently has six players among the top fifty in the world. Russia still has the most, eleven, but their average age is thirty-two, while that of the Chinese players is twenty-five.

The current world champion, Magnus Carlsen, is from Norway and was born in 1990. He has never known a world in which computer chess programs weren't stronger than he is. Ironically, he is very much a "human style" player, whose intuitive positional chess does not directly reflect much silicon influence. This is not the case for many of his contemporaries, however, something we will examine more closely later on.

BEFORE MOVING to my own experiences facing chess machines, it's worth taking a look at the history of this long-running rivalry. Despite my personal investment in such competitions during my career, looking back I can say that the sporting aspect is less interesting than how much we can learn about artificial intelligence and human cognition from the history of computer chess, and especially the competitions between computers and strong humans.

This is not because of how our silicon creations inevitably surpassed us over the board, as much as a holy grail as it was. Nor are many of the games themselves particularly fascinating for nonexperts. The most interesting games are those that represent advances in computer play in some way, because they reflect scientific progress. It's unavoidable that the results will get most of the attention, but it's important to look beyond the wins and losses. In order to use chess as a way to better understand what computers and humans are good at and what they struggle with and why, the moves matter more than the results.

Thanks to the international rating system we use in chess to rank players, a simple chart can show us quite clearly that chess machines have gotten stronger on a steady linear path from the first mainframes to specialized hardware machines to the top programs today. They went from novice level in the 1960s to strong play in the seventies to Grandmaster level in the late eighties and world champion level in the late nineties. There were no giant leaps, just a slow and steady evolution as the global community of developers learned from each other and competed with each other while Moore's law worked its inexorable magic on their hardware.

This growth of machines from chess beginners to Grandmasters is also a progression that is being repeated by countless AI projects around the world. AI products tend to evolve from laughably weak to interesting but feeble, then to artificial but useful, and finally to transcendent and superior to human.

We see this path with speech recognition and speech synthesis, with self-driving cars and trucks, and with virtual assistants like Apple's Siri. There is always a tipping point at which they go from amusing diversions to essential tools. Then there comes another shift, when a tool becomes something more, something more powerful than even its creators had in mind. Often this is the result of a combining of technologies over time, as in the case of the Internet, which is really a half-dozen different layers of technology working together.

It's remarkable how quickly we go from being skeptics to taking a new technology for granted. Despite the rapid pace of technological change that has been the norm for our entire lives, we are briefly amazed, or horrified, or both, by anything new, only to get used to it in just a few years. It's important to keep our heads on straight during that exciting cusp period between shock and acceptance so that we may look ahead clearly and prepare the best we can.

NINE DAYS before I was born in Baku, twenty-two years before I faced thirty-two computers at the same time in Hamburg and thirty-four years before my fateful rematch with Deep Blue, the first recorded game between a chess machine and a human Grandmaster (GM) took place in Moscow. The encounter has largely been forgotten, and it certainly isn't worth being remembered for its chess merits, but it was a landmark nonetheless.

Soviet GM David Bronstein, who passed away in 2006, was a kindred spirit to me in many ways. He was always one of the most curious and experimental minds in chess both on and off the board, and he occasionally ran afoul of the Soviet authorities for his candid nature. Bronstein proposed many innovative ideas for promoting chess and even new variations of the game itself. He was interested in chess computers and artificial intelligence right from the start, and was always eager to play against the newest generation of programs. Bronstein also saw the potential of computer chess to provide insight into how humans think and he wrote many articles on computer chess as his professional playing career wound down.

In 1963, Bronstein was still one of the strongest players in the world, a dozen years removed from drawing a world championship match with the mighty Botvinnik. On April 4, 1963, at the Moscow Institute of Mathematics, he played a full game against a Soviet program running on a Soviet M-20 mainframe computer. I would like to have been able to ask Bronstein what he felt as the first moves were made. He couldn't have been completely sure that the machine played like a beginner. It was a step into the unknown, with no way to prepare for this unique opponent.

It quickly turned out that, to adapt Samuel Johnson's famous quip, the surprise wasn't that the computer played chess well, but that it did it at all. Bronstein played aggressively and toyed with the feeble machine. He allowed the computer to win some material while he moved his pieces into attacking position and flushed out the black king. He finished with a pretty mate in ten, ending the game in just twenty-three moves.

Bronstein's win against M-20 was an urtext of the first generation of (strong) human versus machine chess: the computer gets greedy and is punished. Early programs' evaluation functions were heavily weighted toward material value. That is, which side has more pieces and pawns. It's the easiest factor to evaluate and to program; assign a value to everything on the board and count—and computers are very good at counting. The basic set of values was established two centuries ago: pawns are worth one; knights and bishops are worth three; rooks are worth five; the queen is worth nine.

The king is trickier because, while it's not so powerful in terms of mobility, it must be protected at all costs. The king cannot be captured and if it cannot escape inevitable capture, the game is over: checkmate. One trick is to assign the king a value of one million so the program knows not to put it in danger. Checkmate is an unambiguous and terminal event, another thing computers understand very well. If there is a way to force checkmate in four moves, a computer that looks four moves deep will find it no matter how complicated the position would look to the human eye.

Focusing only on material is also how novice humans play, especially kids. They care only about gobbling up their opponent's pieces and ignore other factors in the position, such as piece activity and whose king is safer. Eventually they learn from experience that, while material is important, it doesn't matter how many of your opponent's pieces you've captured if your king is getting checkmated.

Even the scale of material values is full of exceptions based on the type of position on the board. For example, a well-placed knight can be worth as much or more than a rook with limited scope. During the middlegame—the dynamic, tactical phase of the game—a bishop is likely to be more valuable than three pawns, while the tables can turn in the endgame. Adjusting the various values during a game is possible, but that also adds even more knowledge to the algorithm, slowing down its search.

Early chess machines couldn't learn from experience the way people can. Those greedy kids are learning each time they get checkmated. Even when they lose horribly, they are accumulating useful patterns in their memory. Computers, meanwhile, would make the same mistake over and over, something their human opponents understood and exploited quite well. Even well into the 1980s, if you timed it just right you could replay an entire game against a computer, beating it the same way move for move.

Timing matters because from one microsecond to the next as its search expands, the computer may switch to a different move. A human spending sixty seconds on each move is very unlikely to play much differently than if he spent fifty-five seconds per move, but this isn't true for computers since every sliver of time is put directly into deeper search, with a linear payoff in higher-quality moves.

The apparent similarity between early chess programs and human beginners is a trap, part of the familiar fallacy of expecting computers to think like humans. As Moravec's paradox dictates, computers are very good at chess calculation, which is the part humans have the most trouble with. Computers are poor at recognizing patterns and making analogical evaluations, a human strength. Other than checkmate, nearly every factor that goes into evaluating a chess position is conditional on many other factors. This, along with the slow speed of computers at the time, is why early experts thought it would be impossible to make a strong Type A (brute force) program.

They were wrong, although it would take a while to figure that out. Many of the first programs were attempts at Type B, which sought to intelligently reduce the size of the algorithm's search tree early on the way humans do. Other research groups saw the advantage of tackling the relatively concrete task of improving the machine's search speed and therefore depth, which always improved strength in a predictable way.

The first program that played competent chess was developed at MIT in the late 1950s, a few years ahead of the Soviet program beaten by Bronstein. The Kotok-McCarthy program ran on an IBM 7090 and included some of the techniques that would become the basis for every strong algorithm that followed, including alpha-beta pruning to speed up the search.

The leading Soviet team at the time took a Type A approach, which is interesting considering that they were surrounded by strong chess players, unlike the Americans. Alan Kotok and John McCarthy were both very weak players and had a romantic view of how the game was played. To me, the Soviet embrace of brute force search is not ironic at all, but, to the contrary, reflects a superior understanding of how good chess is played and won. Chess is a very precise game when played well. The advantage of a single pawn is usually more than enough for victory between strong players. Weak players see chess through the lens of their own limitations and frequent mistakes. A novice or nonplayer sees the game as a roller-coaster of cut and thrust, full of blunders on both sides that swing the game this way and that.

If you are designing a chess machine with that romantic vision of the game in mind, scientific precision is less important than moments of inspiration. Occasional blunders aren't so bad if you are counting on your opponent to return the favor, which means there is an element of self-fulfilling prophecy. Type B thinking assumes that the entire system is chaotic and noisy to begin with and just tries to make the best of it by selecting the moves to focus on very early on. Instead of looking at the best twenty moves, or ten, and going from there, the Kotok-McCarthy program started out very narrowly, with just four moves. That is, looking ahead one ply it picked the four best moves and then figured out the three best replies. Then it looked at the two best replies to those moves, etc., getting deeper and narrower.

By design, this is superficially similar to how a strong human player's analysis works, but it ignores that a master's mind can do it effectively only because its assessment of thousands of patterns and the immense parallel processing power of the human brain are choosing that initial menu of three or four candidate moves with formidable accuracy. Expecting a machine to select the right few moves to focus on via calculation, without the benefit of all that experience, is closer to blindfold darts than to blindfold chess.

One of the many handy aspects for chess as an AI laboratory is that we have a good way to measure progress and to test competing theories: over the chessboard! The Soviets started later than the Americans, but their program ITEP had been in development more recently by the time they played a match over telegraph in 1966–67. The ITEP machine, named for the Institute for Theoretical and Experimental Physics in Moscow, was Type A and turned out to be too accurate for the outdated Kotok-McCarthy program and won the match with a score of 3–1.

Around this time, American programmer Richard Greenblatt built on the Kotok-McCarthy concepts with his much better chess understanding, widening the search dramatically. His program Mac Hack VI started with a search width of 15, 15, 9, 9, compared to the Kotok-McCarthy 4, 3, 2, 2. This had the effect of reducing the level of "noise" and making the program far more accurate and stronger. Mac Hack VI also added a database of thousands of opening moves and would become the first computer program to play in a human chess tournament and to receive a chess rating. But despite these improvements and successes, the days of Type B programs were numbered, even more so than those of humans. Brute force was coming.

I WAS INTRODUCED to computers in 1983, although I didn't play chess with them at the time. The British computer company Acorn, the "British Apple," sponsored my match against Viktor Korchnoi in London that year, and of course their products were on display. Businesses, hobbyists, and other early adopters across Europe were paying large sums for the first few generations of home computers and Acorn was doing very well. I won the match, putting me a step away from my first world championship contest with Anatoly Karpov the next year, and was also given an Acorn home computer to take back to Baku. I flew on Aeroflot sitting next to the Soviet ambassador, and my fragile new trophy had its own VIP seat and blanket.

To me, coming from the USSR, owning a computer seemed a little like science fiction. First, I had dedicated my life to climbing up the chess Olympus and this left very little time for other interests. Second, the USSR was still a computing desert outside of research institutions. A Soviet clone of the 1977 Apple II, the AGAT, came out around 1983 and slowly started to appear in schools across the country, but it was far out of reach for most private citizens, costing around twenty times the average monthly Soviet salary. And like most Soviet knock-off tech, it wasn't even a very good clone of a computer that was already six years old. America's *BYTE* magazine wrote in 1984 that "the AGAT wouldn't stand a chance in today's international market, even if they gave it away."

This was far from just a little Cold War jab. The PC revolution was already well under way in America by this time. They were still expensive for what you got, but easily available to the middle class. The hugely popular Commodore 64 was released in August 1982. The standard-setting IBM PC XT came out in early 1983. By late 1984, over

8 percent of American households owned a computer. For comparison, the number of personal computers in Baku, Azerbaijan, a capital city of over a million people, probably went from zero to one when the plane landed with me and my Acorn.

I would like to say this first encounter with a computer was a transformative moment, but as I said, I was a little busy at the time. My cousins and friends mostly used my eight-bit Acorn, a BBC Micro model, I believe, to play video games. One in particular would come to alter my perception of computers and my life in an important way, but it wasn't a chess game. It involved moving a little green frog across traffic.

One day early in 1985, I received a package from a stranger named Frederic Friedel, a chess fan and science writer based in Hamburg, Germany. He sent me a nice note and a floppy disk containing several computer games, including my new favorite, called Hopper. I admit I spent much of my free time over the next few weeks playing Hopper and setting ever-higher record scores.

A few months later, I traveled to Hamburg for several events, including the computer simul, and I also visited Mr. Friedel at his suburban home. I met his wife and two young sons, Martin, age ten, and Tommy, age three. They made me feel quite at home and Frederic was eager to show me the latest developments on his own computer. I managed to work into the conversation that I had completely mastered one of the little games he had sent me.

"You know, I'm the best Hopper player in Baku," I said, omitting any mention of the total lack of competition. I told him that I had scored sixteen thousand points and was a little surprised that this extraordinary number failed to elicit at least a raised eyebrow.

"Very impressive," Frederic said, "but that's not such a big score in this house."

"What? You can beat it?" I asked.

"No, not me."

"Ah, okay, Martin must be the video game whiz."

"No, not Martin."

It was with a sinking feeling that I realized the smile on Frederic's face meant that the household Hopper champion was the three-year-old. I was incredulous. "You can't mean Tommy!" My fears were confirmed when Frederic led his little boy over to the computer and sat him down next to us as the game loaded. Since I was the guest they let me go first and I rose to the occasion with a personal best of nineteen thousand points.

My success was short-lived, however, as Tommy took his turn. His little fingers were a blur and it wasn't long before the score read twenty thousand, then thirty thousand. I conceded defeat to avoid having to sit watching through dinnertime.

Losing to a little kid at Hopper was easier on my ego than any loss to Karpov, but it still gave me food for thought. How was my country going to compete with a generation of little computer geniuses being raised in the West? Here I was, one of the few people in a major Soviet city with a computer, and I had been handily outperformed by a German toddler.

And so, when I signed a sponsorship deal with the computer company Atari in 1986, I took as payment over fifty of their newest machines to bring back to form a youth computer club in Moscow, the first of its kind in the Soviet Union. I continued to supply the club with hardware and software acquired on my travels and it became a hub for many talented scientists and hobbyists.

They would often give me lists of equipment they wanted for their projects, leading to some amusing scenes at the airport when I would return from my travels like Father Christmas delivering presents. Mixed in with the chess fans welcoming me home, there would be computer experts hoping I'd managed to find the items on their wish lists. I even recall being met by a shout that would get quite a lot of attention from security at any airport today: "Garry! Did you bring the Winchester?!" It was a much-coveted type of hard drive.

Frederic and I also had the chance to talk about the potential implications computers had for professional chess. Businesses were rapidly adopting PCs for spreadsheets, word processing, and databases, so why couldn't this sort of Hoppering be done for chess games? This would be a powerful weapon, one that I couldn't afford to be the last to have.

As described above, our conversations led to the creation of the first version of ChessBase, a name that soon became synonymous with professional chess software. In January 1987, I tried out an early version of the program to prepare for a special simultaneous exhibition against a strong team. I had narrowly lost a similar event in 1985, playing against eight members of a professional German league team at the same time. I had come in tired and overconfident, especially since I didn't know much about most of my opponents and had no way to quickly prepare for them.

For the rematch, I discovered how much ChessBase was going to change professional chess and my life. Using an Atari ST and a ChessBase diskette labeled "00001" that I was given by Frederic and Matthias, I was able to bring up and review my opponents' previous games in hours, a process that would have taken weeks without a computer. With just two days of preparation I felt comfortable going into the match and won in crushing fashion, 7–1. That was when I knew I was going to be spending a lot of time in front of a computer for the rest of my career. I just didn't realize yet how much of that time would be spent playing against them.

HOW QUICKLY and completely computers came to dominate chess preparation was illustrated a few years later when an interviewer and photographer came to where I was staying. The photographer wanted some pictures of me at a chessboard to accompany the story. The only problem? I didn't have a chessboard with me! All my preparation was done on my laptop, a Compaq that really stretched the definition of "portable." It must have weighed close to twelve pounds. Even so, it was far lighter and more efficient than traveling with my paper notebooks and a stack of opening encyclopedias. The advantages would accumulate when the Internet made it possible to download the latest games nearly as soon as they had been played, instead of having to wait weeks or months for them to be published in a magazine.

Soon nearly every Grandmaster traveled to every tournament with a laptop, although there was a jagged generational break in this regard.

Many older players found them too complicated, too alien, especially after having decades of success with their traditional training and preparation methods. Laptops were also still very expensive, and few players had my advantages of sponsorship deals and world championship prize purses.

How professional chess changed when computers and databases arrived is a useful metaphor for how new technology is adopted across industries and societies in general. It's a well-established phenomenon, but I feel that the motivations are underanalyzed. Being young and less set in our ways definitely makes us more open to trying new things. But simply being older isn't the only factor that works against this openness—there is also being successful. When you have had success, when the status quo favors you, it becomes very hard to voluntarily change your ways.

In my lectures to business audiences I call this the "gravity of past success," and often give a painful example from my own career: the loss of my world championship title to Vladimir Kramnik in 2000. I was at the height of success at the time, in the midst of an unprecedented winning streak at top-level tournaments and raising my rating to its highest peak ever. I felt great and had prepared deeply for our October match in London, scheduled for sixteen games. Kramnik was my most dangerous opponent, twelve years younger and with years of strong performances against me. But it was his first world championship match and my seventh. I had experience, better results, and felt good. How could I lose?

The answer was "by playing into my opponent's strength and refusing to adapt." Kramnik had prepared very cleverly, using his turns with the black pieces to draw me into tedious positions I disliked. This was entirely to his credit, and it was up to me to find a strategic response for the rest of the match. But instead of avoiding these positions entirely and playing to my strengths, I continued to charge straight ahead like a bull at a red cape. I eventually lost the match with two losses, thirteen draws, and without winning a single game.

I was thirty-seven at the time, not exactly ancient. And I was never afraid of pushing myself to stay on the cutting edge, including my embrace of technology. My weakness was a refusal to admit that Kramnik had outprepared me—preparation was supposed to be *my* strong suit. Each one of my successes up to that moment was like being dipped in bronze over and over, each success, each layer, making me more rigid and unable to change, and, more importantly, unable to see the need to change.

This metaphorical gravity isn't only a problem for individuals, or only a matter of ego. Fighting against disruption and change is also a standard business practice, one that is usually employed by a market leader trying to protect that lead. There are countless examples of this from the real world, but I'll take one ad absurdum case from science fiction, the 1951 movie *The Man in the White Suit*, starring Alec Guinness. Guinness, the protagonist, is a rogue research chemist who invents a miracle fiber that never wears out and never gets dirty. Instead of the fame, riches, and Nobel Prize you might expect, he ends up being chased through the streets by angry mobs once various interest groups realize what his invention will mean. No more demand for new cloth, so the textile industry will be wiped out along with thousands of union jobs. No more need for laundry soap or laundry workers, who join in the pursuit.

Far-fetched? Certainly, but I don't think you have to have my suspicious mind to wonder if lightbulb companies would sell an indestructible and everlasting bulb if they could make one. But resisting change and delaying it to squeeze a few more dollars out of an existing business model usually just makes the inevitable fall all the worse. I once made a television commercial for the search engine company AltaVista in 1999, but that didn't mean I wanted to follow it to oblivion when the chess equivalent of Google came along.

I was in my twenties when the digital information wave rolled over the chess world, and it was a fairly gradual one, not a tsunami. Flicking through games on a screen was far more efficient than on printed materials, a real competitive advantage but not a nuclear bomb. The impact of the Internet a few years later was just as great, dramatically accelerating the information warfare that Grandmasters wage against each other over the board. A brilliant new opening idea played in a game in Moscow on Tuesday could be imitated by a dozen players around the world on Wednesday. It shortened the lifespan of these secret weapons, what we call opening novelties, from weeks or months to hours. No more could you hope to ensnare more than a single opponent with a clever trap.

Of course, that was only true if your opponents were also online and up to date, which wasn't the case for a while. Asking a fifty-year-old Grandmaster to ditch his beloved leather-bound notebooks of analysis, printed tournament bulletins, and other preparation habits was like asking a successful writer to switch to a word processor or an artist to start drawing on a screen instead of a canvas. But in chess, it was a matter of adapting to survive. Those who quickly mastered the new methods thrived; the few who didn't mostly dropped down the rating lists.

There's no way to prove causation, but I'm certain that the rapid decline of many veteran players in the 1989–95 span, when ChessBase became standard, had much to do with their inability to adjust to the new technology. The 1990 rating list included over twenty active players born before 1950 among the top one hundred in the world. By 1995, there were just seven, and only one among the elite: the ageless Viktor Korchnoi, born in 1931, who was my opponent in that 1983 London candidates match sponsored by Acorn. Another exception was my great rival Karpov, born in 1951, who stayed near the top into his fifties despite his personal reluctance to embrace computers and the Internet. But along with his tremendous talent and experience, as a former world champion with considerable resources he also counted on the assistance of colleagues for his research, an advantage few others had. Reducing the advantage of being able to afford assistants, or "seconds," as they are called in chess, in a tribute to the age of duels, was one of the many democratizing impacts technology had on the chess world.

While they may have shortened the careers of a few older players, computers also enabled younger players to rise more quickly. Not just the playing engines, but because of how PC database programs allowed elastic young brains to be plugged into the fire hose of information that was suddenly available. Even I am startled to watch kids zipping from one game to the next, one branch of analysis to another, in the blink of an eye. Computer-centric training also has drawbacks, and I'll get to those later, but there is no doubt it tipped the playing field, or the chessboard, even further toward youth. As my professional career progressed, not only would I be facing the challenge of every champion to fend off the next generation of players, but it would be a generation that had grown up with sophisticated tools that hadn't existed when I was a kid.

I was born just in time to ride this wave instead of being swept away by it. But this timing also put me on the front lines against a new enemy that was growing stronger by the day. The chess machines were finally coming for the world champion, and, as of November 9, 1985, that was me.

WHEN WILL a chess-playing machine be able to beat the world champion?" This question was put to every chess programmer in history dozens of times. As you could expect, the earliest predictions, from the days of the digital computer's infancy, were wildly off the mark. At least the Carnegie Mellon group's daring 1957 promise of 1967 was in some ways avenged since it was a group from the same school whose Deep Blue eventually did the job—if forty years later instead of ten.

At the twelfth annual North American Computer Chess Championship, held in Los Angeles in 1982, the world's best chess machines battled each other for supremacy. Ken Thompson's special-purpose hardware machine Belle continued to show its superiority over the rest, and to show the potential for a hardware architecture and customized chess chips that would later be realized by Deep Blue. Thompson, with Belle codeveloper Joe Condon, worked at the famous Bell Laboratories and, among many other accomplishments, was one of the creators of the Unix operating system.

As far as results go, Belle was the definitive answer to the dilemma Claude Shannon presented in 1950 between "fast but dumb" Type A brute force and "smart but slow" Type B artificial intelligence programs. It was now clear the brute force, with a fast-enough search, was enough to play very strong chess. Despite a relative lack of knowledge and other evaluation limitations, Belle's raw speed, up to 160,000 positions per second, produced results that were leaving smarter microprocessor machines and even Cray supercomputers in the dust. Interviews with various computer chess luminaries at the 1982 event about when a machine would defeat the world champion (then Karpov) revealed cautious optimism.

Monty Newborn, long one of the motive forces behind computer chess, especially as a promoter and organizer, was remarkably optimistic with his answer of five years. Another expert, Mike Valvo, who was also an International Master, said ten years. The creators of the popular PC program Sargon nailed it exactly with fifteen. Thompson thought it was still twenty years off, putting him on the pessimistic side of the vast majority that said it would happen around the year 2000. A few even said it would never happen, reflecting some of the problems even the faster machines were having with the law of diminishing returns that arose when adding chess knowledge to their creations. But this was the last time that the question was "when or whether" instead of just "when."

By the late 1980s, after another decade of steady progress, the computer chess community was well aware that time was on their side in the human versus machine contest and they could narrow the range of their prognostications effectively. In 1989, at the World Computer Chess Championship in Edmonton, Canada, a survey of the forty-three experts present reflected the recent achievements in human-machine play. A computer had just beaten a Grandmaster in tournament play for the first time the year before and the road map of further improvement was coming into sharp focus: a little more knowledge and a lot more speed. Still, only one correctly picked 1997 as the year of destiny, while other guesses ranged from within a decade of that. Notable was a member of the Deep Blue team, Murray Campbell, guessing 1995, and Claude Shannon himself saying 1999.

It's a little unfair to highlight the early erroneous forecasts and spurious rationales that came out of the computer chess community over the years. After all, human calculation may be weak, but our hindsight is always perfect. But there is a point to it, since in many cases their sins, both their overoptimism or resigned pessimism, are a distant mirror for today's flood of predictions about artificial intelligence.

Overestimating the potential upside of every new sign of tech progress is as common as downplaying the downsides. It's easy to let our imaginations run wild with how any new development is going to change everything practically overnight. The unforeseen technical roadblocks that inevitably spring up are only one reason for this consistent miscalculation. Human nature is simply out of sync with the nature of technological development. We see progress as linear, a straight line of improvement. In reality, this is only true with mature technologies that have already been developed and deployed. For example, the way Moore's law accurately described the advances in semiconductors, or the way solar cell efficiency is improving at a slow but steady pace.

Before that predictable progress phase, there are two previous phases: struggle and then breakthrough. This fits the axiom of Bill Gates, "We always overestimate the change that will occur in the next two years and underestimate the change that will occur in the next ten." We expect linear progress, but what we get are years of setbacks and maturation. Then the right technologies combine or a critical mass is reached and boom, it takes off vertically for a while, surprising us again, until it reaches the mature phase and levels off. Our minds see tech progress as a straight diagonal line, but it's usually more of an S-shape.

The chess machines of the fifties and sixties were still in the struggle phase. Researchers were doing a lot of experimentation with few resources, still trying to figure out if Type A or Type B was the most promising, all while using primitive coding tools on hardware that was incredibly slow. Was chess knowledge the key? Was speed the most important factor? With so many of the basic concepts still up in the air, each new breakthrough felt like it could be the big one.

One strong chess player decided he could put the scientists' optimism to good personal advantage. Long before I took my turn as the computer chess world's "most wanted," a Scottish International Master named David Levy turned beating them into a profitable sideline.

In 1968, after hearing two eminent AI experts predict that a machine would beat the world champion in a decade, Levy made a famous bet that no computer would be able to defeat him in a match in that span. If you looked at the modest progress chess machines made in the two decades after Claude Shannon created the road map in 1949, you could see his point.

(To quickly clarify some terminology, at around 2400 in rating, an International Master is above master (2200) and a rank below Grandmaster (2500 and up). Twenty-seven hundred is considered elite today, with around forty players in the world surpassing that mark, on up to Magnus Carlsen's record of 2882. My peak was 2851 in 1999 and my rating was 2795 when I played my second match with Deep Blue. It should be noted that ratings are getting higher over time: Bobby Fischer's 1972 peak of 2785 was like Mount Everest in its day, but quite a few players have surpassed that number, while I cannot say they have surpassed Fischer. We say "match" for a series of games between two opponents, as opposed to a tournament with many players.)

Levy was much stronger than the programs of the early 1970s; no program would approach master level until the bet came due. What's more, Levy was also very savvy about the strengths and weaknesses of computer chess players. He understood that while they were getting quite dangerous in tactical complications thanks to their increasingly deep search abilities, they were clueless about strategic plans and the subtleties of endgame play. He would maneuver patiently, employing an anti-computer strategy of "doing nothing, but doing it well" until the machine would overextend and create weaknesses in its own position. Then Levy would clean up on the board—and in his bets.

It looked like smooth sailing for Levy until the appearance of a program from Northwestern University, called simply "Chess." The program by Larry Atkin and David Slate was the first chess machine to play the strong, consistent chess needed to beat experts without a serious human blunder. By 1976, version 4.5 of Chess was good enough to win a weak human tournament. The next year, 4.6 finished first in an open tournament in Minnesota, reaching a rating performance near expert level, if not quite master.

The struggle phase of development was over and the rapid growth phase had begun. The combination of faster hardware and twenty years' worth of programming improvements had crested. After decades of disappointments after overvaluing potential advances, real progress came faster than anyone expected. When the time came for Levy to meet the computer world champion in 1978, Chess 4.7 was far stronger than he imagined any machine would be by that time. It wasn't quite strong enough, however, although it did score a draw and a win against him in their six-game match.

Levy went on to be an important force in the machine chess world and has written countless books and articles on the subject. He's the president of the International Computer Games Association (ICGA), the organization that oversaw my 2003 match against the program Deep Junior in New York City. In 1986, Levy wrote an article in the *ICGA Journal* titled "When Will Brute Force Programs Beat Kasparov?" I think he was quite glad to have the target on his back transferred to someone else.

Levy collected his winnings and threw down a new gauntlet, a bounty of $1,000 to the computer that could beat him. The American science magazine *Omni* sweetened the pot with another $4,000. It would be another ten years before someone collected that money, a group of grad students at Carnegie Mellon with a custom hardware–based chess machine called Deep Thought.

CHAPTER 4

WHAT MATTERS TO A MACHINE?

> "All right," said Deep Thought. "The Answer to the Great Question . . ."
>
> "Yes . . .!"
>
> "Of Life, the Universe and Everything . . ." said Deep Thought.
>
> "Yes . . . !"
>
> "Is . . ." said Deep Thought, and paused.
>
> "Yes . . . !"
>
> "Is . . ."
>
> "Yes . . . !!! . . .?"
>
> "Forty-two," said Deep Thought, with infinite majesty and calm.
>
> "Forty-two!" yelled Loonquawl. "Is that all you've got to show for seven and a half million years' work?"
>
> "I checked it very thoroughly," said the computer, "and that quite definitely is the answer. I think the problem, to be quite honest with you, is that you've never actually known what the question is."

As with all the best jokes, there is profundity in this conversation between the universe's fastest computer and its makers from Douglas Adams's *The Hitchhiker's Guide to the Galaxy.* We are forever looking for answers without first making sure we understand the questions, or if they are the right ones. In my lectures on the human-machine relationship, I'm fond of citing Pablo Picasso, who said in an interview, "Computers are useless. They can only give you answers." An answer means an end, a full stop, and to Picasso there was never an end, only new questions to explore. Computers are excellent tools for producing answers, but they don't know how to ask questions, at least not in the sense humans do.

In 2014, I got an interesting response to this assertion. I had been invited to speak at the headquarters of the world's largest hedge fund in Connecticut, Bridgewater Associates. In a revealing turn of events, they had hired Dave Ferrucci, one of the creators of the IBM artificial intelligence project Watson, famous for its triumphs on the American television quiz show *Jeopardy*. Ferrucci sounded disillusioned by IBM's focus on a data-driven approach to AI, and how it wanted to exploit the impressive Watson and its sudden celebrity by turning it into a commercial product as quickly as possible. He had been working on more sophisticated "paths" that aimed at explaining the "why" of things, not only finding useful correlations via data mining. That is, he wanted AI to investigate beyond immediate practical results, and for those results to be revealing instead of simply an answer.

Interestingly, Ferrucci decided that the famously iconoclastic Bridgewater could be the place to do this sort of ambitious experimental research instead of IBM, one of the world's largest technology companies. Of course, first and foremost, Bridgewater was looking for predictive and analytical models to improve their investment results. They believed it was worthwhile to back Ferrucci's attempts to, as he put it, "imagine a machine that can combine deductive and inductive processes to develop, apply, refine and explain a fundamental economic theory."

This is a grail worthy of a holy quest, especially "explain." Even the strongest chess programs in the world can't explain any rationales behind their brilliant moves beyond elementary tactical sequences. They play a strong move simply because it was evaluated to be better than everything else, not by using the type of applied reasoning a human would understand. It's still very useful to have a super-strong machine to play against and to analyze with, of course, but for a nonexpert it can be a bit like asking a calculator to be your algebra tutor.

Ferrucci's interjection during my lecture cut to the core of the matter as effectively as Picasso and Douglas Adams. He said, "Computers *do* know how to ask questions. They just don't know which ones are important." I love this because it has several layers of meaning, and all of them provide useful insight.

First, we can take it literally. The simplest program can ask you a scripted question and record the answer. This isn't AI in any of its many definitions, however; it's just automated digital note-taking. Even if the machine asks in a realistic voice and follows up your answers with appropriate questions, it's probably doing little more than the most primitive data analysis. This sort of thing has been a common help feature in software and on websites for over a decade, if without the natural voice component. You type in your question or problem and the help system or chatbot picks out the key words—"crash," "audio," "PowerPoint"—and then offers you some related help pages and follow-up questions it thinks are related.

Anyone who has used a search engine like Google has experience with these systems, which means pretty much everyone. Most of us realized long ago that there is no point to googling "What is the capital of Wyoming?" when simply "capital Wyoming" produces the same results with less effort. People prefer to use more natural language when speaking compared to typing, however, and like to say full sentences when talking to Siri, Alexa, Ok Google, Cortana, and the other virtual assistants that are increasingly listening to our every word. This is one reason why there is such a push now in social robotics, one of the terms used for studying how people interact with artificially intelligent technology. How our robots look, sound, and behave is a big part of how we choose to use them.

When I spoke at a social robotics conference in Oxford in September 2016, I chatted with fellow presenter Dr. Nigel Crook and his robot, Artie. Dr. Crook works in AI and social robotics at Oxford Brookes University and he emphasized how critical it is to study the use of robots in public spaces, where people are equal parts fascinated and afraid of them. A disembodied voice from your phone is one thing, but it's quite another when it is emanating from a mechanical face and body. Regardless how you feel about them, there are going to be many more of them everywhere you go.

Getting back to whether or not computers can ask questions in the deeper sense that AI visionaries like Ferrucci are working on, more sophisticated algorithms are being developed to investigate the motivations and causations of events in data, not merely ranking correlations to answer search and trivia questions. But to know which questions are the right questions, you have to know what is important, what *matters*. And you cannot know that unless you know which outcome is most desirable.

I speak regularly about the difference between strategy and tactics, and why it's essential to first understand your long-term goals so you don't confuse them with reactions, opportunities, or mere milestones. The difficulty of doing this is why even small companies need mission statements and regular checkups to make sure they are staying on course. Adapting to circumstances is important, but if you change your strategy all the time you don't really have one. We humans have enough trouble figuring out what we want and how best to achieve it, so it's no wonder we have trouble getting machines to look at the big picture.

Machines have no independent way to know if or why some results matter more than others unless they've been programmed with explicit parameters or have enough information to figure it out on their own. What does it even mean to say something matters to a machine? Either a result is significant or it's not, based on what it has been told is significant, and humans have to establish these values for them. At least, that's the way it has been for a long time. But our machines are starting to move from surprising us with results to surprising us with the methods they use to find results, and that is a huge difference.

To use a simplified example, a traditional chess program knows the rules of the game. It knows how the pieces move and how checkmate works. It is also programmed with the values of the pieces (one for pawn, nine for queen, etc.) and other knowledge like piece mobility and pawn structure. Anything that goes beyond the rules is classified as knowledge. If you program it with the knowledge that a pawn is worth more than a queen it will go into battle throwing the queen and the game away with no hesitation.

But what if you don't provide that knowledge to the machine at all? What if you only tell it the rules and let it figure out the rest on its own? Let the machine figure out that rooks are more valuable than bishops, that doubled pawns can be weak, that open files can be useful. This opens up the possibility of not only creating a strong chess machine, but also that humans will learn something new from what the machine discovers and how it discovers it.

This is what different systems are indeed doing today, using techniques like genetic algorithms and neural nets to basically program themselves. Unfortunately, they have not proved to be stronger than the traditional fast-searching programs that rely more on hard-coded human knowledge—at least not yet. But this is the fault of chess, not of the methods. The more complex the subject, the more likely it is to benefit from an open, self-creating algorithm versus fixed human knowledge. Chess just isn't complex enough and even I can admit that there is more to life than chess.

It took thirty years, but my beloved game was revealed to be too vulnerable to brute force fast searching to require strategic thinking from machines in order to defeat the best humans. As much work as went into tuning Deep Blue's evaluation function and training its openings, the depressing truth is that a few years and a new generation of faster chips later, none of it would have mattered very much. For better or worse, chess just wasn't deep enough to force the chess-machine community to find a solution beyond speed, something many among them lamented.

In a 1989 article, two of the leading figures in computer chess wrote an essay titled "Perspectives on Falling from Grace" that critiqued the methods by which chess machines had finally approached Grandmaster strength. Soviet computer scientist Mikhail Donskoy was one of the creators of the Kaissa program that won the first computer chess world championship in 1974. Jonathan Schaeffer of Canada and his colleagues at the University of Alberta have been at the forefront of game-playing machines for decades. Along with his work on chess, he's produced a strong poker program and his program Chinook battled for the checkers world championship and nearly solved the game entirely.

In their provocative essay in a leading computer chess journal, Donskoy and Schaeffer describe how computer chess had separated from AI over the years. They credit this separation to being a consequence of the overwhelming success of the alpha-beta search algorithm. Why look at anything else if the winning method was already in hand? As they write, "It is unfortunate that computer chess was given such a powerful idea so early in its formative stages." Winning was what mattered and faster was better, so engineering took over from science. Patterns, knowledge, and other humanlike methods were discarded as the super-fast brute force machines took home all the trophies.

For many, this was a huge blow. Chess had been an important research topic in psychology and cognition practically since the discipline was created. In 1892, Alfred Binet studied chess players as a continuation of his research on "mathematical prodigies and human calculators." His results had a major impact on the study of different types of memory and mental performance. Binet's insights into the differences between innate talent and acquired knowledge and experience defined the field. "One becomes a good player," he wrote. "But one is born an excellent player." Binet would go on to create the IQ test with Theodore Simon. In 1946, Binet's work was advanced by the Dutch psychologist Adriaan de Groot, whose extensive testing of chess players revealed the importance of pattern recognition and peeled away at the mysteries of human intuition in decision making.

John McCarthy, the American computer scientist who coined the term "artificial intelligence" in 1956, called chess the "*Drosophila* of AI," referring to how the humble fruit fly was the ideal subject for countless seminal scientific experiments in biology, especially genetics. By the end of the 1980s, the computer chess community had largely resigned this great experiment. In 1990, Ken Thompson of Belle was openly recommending the game Go as a more promising target for real advances in machine cognition. In the same year, the compendium *Computers, Chess, and Cognition* included an entire section on Go, titled "A New *Drosophila* for AI?"

The nineteen-by-nineteen Go board with its 361 black and white stones is too big of a matrix to crack by brute force, too subtle to be decided by the tactical blunders that define human losses to computers at chess. In that 1990 article on Go as a new target for AI, a team of Go programmers said they were roughly twenty years behind chess. This turned out to be remarkably accurate. In 2016, nineteen years after my loss to Deep Blue, the Google-backed AI project DeepMind and its Go-playing offshoot AlphaGo defeated the world's top Go player, Lee Sedol. More importantly, and also as predicted, the methods used to create AlphaGo were more interesting as an AI project than anything that had produced the top chess machines. It uses machine learning and neural networks to teach itself how to play better, as well as other sophisticated techniques beyond the usual alpha-beta search. Deep Blue was the end; AlphaGo is a beginning.

THE LIMITATIONS of chess weren't the only fundamental misconceptions in the equation. The founding computer science aspect of AI also revealed its limitations. The basic suppositions behind Alan Turing's dreams of artificial intelligence were that the human brain is itself a kind of computer and that the goal was to create a machine that successfully imitated human behavior. This concept has been dominant for generations of computer scientists. It's a tempting analogy—neurons as switches, cortexes as memory banks, etc. But there is a shortage of biological evidence for this parallel beyond the metaphorical and it is a distraction from what makes human thinking so different from machine thinking.

The terms I prefer to highlight these differences are "understanding" and "purpose." I will begin with the first. A machine like Watson that is designed to understand natural human language has to sort through millions of clues to establish enough context to make sense of something that is instantly obvious to a human. The simple sentence "the chicken is too hot to eat" can mean that a barnyard animal is ill or that dinner needs to cool down. There is no chance of a human mistaking the speaker's meaning despite the inherent ambiguity of the sentence itself. The context in which someone would say this would make the meaning obvious.

Applying context comes naturally to humans; it's one way our brains handle so much data without having to consciously figure things out constantly. Our brain does the work in the background without any noticeable effort on our part, nearly as effortlessly as breathing. A strong chess player knows in a glance that a certain type of move is good in a certain type of position and you know you will enjoy a pastry that looks a particular way. Of course, these background intuition processes are sometimes wrong, leaving you with a lost position or a second-rate snack, and as a result your conscious mind will probably assert itself a little more next time you are in that situation and second-guess your intuition.

In contrast, a machine intelligence has to build context for every new piece of data it encounters. It has to process a huge amount of data to simulate understanding. Imagine all the questions a computer would have to answer before being able to diagnose the problem with our hot chicken. What is a chicken? Is the chicken alive or dead? Are you on a farm? Is a chicken something you eat? What is eating? When I used this example at a lecture to an audience that was mostly English second-language, someone pointed out later that there was even an additional element of ambiguity because "hot" in English can mean the level of spice or the temperature of food.

Despite all this complexity even in simple sentences, Watson showed it is possible for a machine to provide accurate answers if there is enough relevant data available and it can be accessed quickly enough and cleverly enough. As with a chess engine crunching through billions of positions to find the best move, language can be broken down into values and probabilities to produce a response. The faster the machine, the more and better quality the data, and the smarter the code, the more accurate the response is likely be.

Adding a bit of irony regarding whether or not computers can ask questions, the format of the television game show *Jeopardy*, where Watson showed off its capabilities by defeating two human former champions, requires contestants to provide their answers in the form of a question. That is, if the show's host says, "This Soviet program won the first World Computer Chess Championship in 1974," the player would press the buzzer and answer, "What was Kaissa?" But this odd convention is simple protocol with no bearing on the machine's ability to find the answers in its fifteen petabytes of data.

Regardless, the output is sufficient. The performance is better than that of a human. There is no understanding, but there was never intended to be any. A medical diagnostic AI can dig through years of data about cancer or diabetes patients and find correlations between various characteristics, habits, or symptoms in order to aid in preventing or diagnosing the disease. Does it matter that none of it "matters" to the machine as long as it's a useful tool?

Perhaps not, but it matters very much to those who want to build the next generation of intelligent machines, machines that learn for themselves faster than we could possibly teach them. Humans don't learn a native language from grammar books, after all.

The trajectory so far has been as follows: We create a machine that follows strict rules in order to imitate human performance. Its performance is poor and artificial. With generations of optimization and speed gains, performance improves. The next jump occurs when the programmers loosen the rules and allow the machine to figure out more things on its own, and to shape or even ignore the old rules. To become good at anything you have to know how to apply basic principles. To become great at it, you have to know when to violate those principles. This isn't only a theory; it's also the story of my own battles against chess machines over two decades.

CHAPTER 5

WHAT MAKES A MIND

ONE OF THE PROBLEMS with all the predictions and statistics-based estimates on machine progress in chess is that chess is a competitive sport. And, no, I won't be deflected into pointless arguments about whether or not chess is a sport, or a game or a hobby or an art or a science either. I am not going to argue with the International Olympic Committee, which has rejected petitions from bridge and chess to become Olympic sports on the grounds that "mind sports" don't require physical prowess in the act of performing the discipline, although anyone who has watched chess masters bash out dozens of moves with seconds on the clock might disagree.

It is simply chess, and, at least when it is played competitively, it contains most of the elements that define all sports. The most critical of these sporting elements when it comes to human chess versus machine chess is that it is a competition. The goal isn't to play well; that is only the means to the end of winning the game. We can talk about seeking truth over the board and about finding artistic fulfillment, but at the end of the long day at the board, it's win, lose, or draw.

Another aspect of chess as a sport is the intense psychological and physiological exertion involved in a competitive chess game, and the crisis after the game. What sports science calls the "stress response process" is at least as powerful in chess as it is in more physical sports. When I say exertion, I am not referring only to the mental gymnastics of moving the pieces in our minds, but also the huge nervous tension that fills you before and during the game, tension that rises and falls with every move and with every idea that passes through your mind while at the board. This tension lasts for hours and a balanced game is a roller coaster of emotions as fortunes change and the battlefield shifts. Delight can give way to depression in an instant and reverse again a move later, leaving even the most sanguine player exhausted from adrenaline. Managing this nervous energy during each game, and during the ups and downs of an event that may last weeks, is an essential skill for a Grandmaster.

Recovery is no small matter, especially from defeat. There are no convenient deflections to share the blame for a loss in chess. There are no referees to blame, no sun in your eyes, or teammates to let you down. There is no luck factor as you have with cards or dice. If you lose it is because the other player beat you, because you failed. Every competitive person has to have a sizable ego, so losses can hit particularly hard in chess. There must also be a critical balance between putting a bad loss out of your mind so you can go into your next game full of the confidence you need and being able to objectively analyze your failures so you do not repeat them.

Chess is also a sport in how imperfectly it is played, especially by humans but, still, also by machines. In 2003, I started a series of chess books called *My Great Predecessors* that included my analysis of hundreds of classic games from the greatest players. Even with computer-aided analysis, many of these masterpieces lived up to their reputations as tremendous achievements. But even these legendary games between our greatest champions were often riddled with mistakes and inaccuracies. It was humbling to find this was also the case when I put my own games under the microscope in my *Modern Chess* series a few years later. The saying that the victor is the one who makes the next to last mistake is very true. But, as another saying goes, this is a feature of chess, not a bug. If you make a relatively minor error and fall into a difficult position you can hope your opponent will falter in return, especially if you put up a stout defense.

German world champion Emanuel Lasker was the greatest proponent of chess as a pitched battle. Lasker was a philosopher and mathematician from the days when chess was still a gentleman's club pastime and whose biography was prefaced by his peer and admirer, Albert Einstein. Lasker employed psychology and knowledge of his opponent as capably as he applied chess acumen, holding the title for a record twenty-seven years. In his 1910 book, *Common Sense in Chess*, Lasker made this statement before moving on to how to improve the reader's opening play:

> Chess has been represented, or shall I say misrepresented, as a game—that is, a thing which could not well serve a serious purpose, solely created for the enjoyment of an empty hour. If it were a game only, Chess would never have survived the serious trials to which it has, during the long time of its existence, been often subjected. By some ardent enthusiasts Chess has been elevated into a science or an art. It is neither; but its principal characteristic seems to be—what human nature mostly delights in—a fight.

Lasker was a pioneer in the psychological approach to chess, writing that the best move was the one that made your opponent most uncomfortable. That is, "to play the man, not the board." Of course, strong moves can disturb any player, but Lasker made it clear that certain types of moves and strategies were stronger against different players. His idea of objective truth on the chessboard was that winning was everything and that understanding the good and bad qualities of your opponent was essential to winning.

Lasker's approach was quite a break from his world champion predecessor, Wilhelm Steinitz. A proud dogmatist, Steinitz said that he would never consider his opponent's personality. "So far as I am concerned my opponent might as well be an abstraction or an automaton." Fateful words. As he said this in 1894, Steinitz never had to test this theory against an actual automaton. I was not to be so fortunate.

The point of this brief excursion into the competitive and psychological aspects of chess is that all of it is meaningless when you play against a computer. Well, not entirely because you still have to navigate these factors yourself, but it's meaningless to the machine. A machine won't get overconfident when it has a superior position or dejected when it is worse. A computer won't tire during a tense six-hour battle, get nervous as its clock ticks down, or get hungry, or distracted, or need restroom breaks. Worse, knowing your opponent is immune makes it even harder to properly navigate your own nervous system when facing a machine.

It's a very strange feeling. So much of the experience is the same as any other game: the board, the pieces, the opponent sitting across from you. But this opponent is only a human puppet, relaying the moves of an algorithm. If chess is a war game, how can you motivate yourself to go to war against a piece of hardware?

THIS ISN'T an idle question about psychology; motivation matters very much. The ability to maintain an intense level of concentration for an extended period of time is a significant part of elite chess performance. The "chess talent" that psychologists like Binet and de Groot searched for has an ineffable nature akin to astronomical phenomena that can only be indirectly observed by their effects. Until more sophisticated tests or scans reveal our secrets, we know only that such talent exists because some players are much better than others, and that this disparity goes well beyond what can be explained by experience or training.

The science writer Malcolm Gladwell famously formulated a "ten thousand hours" theory in his book *Outliers* that says practice, not innate talent, is what makes the difference in exceptional human achievement. When challenged with the obvious fact that Kenyan distance runners and Jamaican sprinters aren't just practicing more than everyone else, Gladwell responded in the *New Yorker* by explaining that the theory applied only to "cognitively complex activities," concluding, "In cognitively demanding fields there are no naturals." He even dedicates a paragraph to chess, and how many hours various prodigies studied before reaching the master or Grandmaster level.

Gladwell later clarified further in a Q&A on the website Reddit, writing that practice alone wasn't enough, and that "I could play chess for 100 years and I'll never be a Grandmaster. The point is simply that natural ability requires a huge investment of time in order to be made manifest." I cannot disagree with this statement in isolation, being the product of its truth myself. If the bar is set to the Grandmaster level, the huge amount of empirical knowledge required for the opening and endgame phases alone makes extensive study and practice a necessity. And the thousands of tactical and positional patterns that Grandmasters recognize so effectively can only be acquired by experience.

But while Gladwell isn't denying that cognitive talents exist, he is underestimating their potency, especially at the early stages of development. Saying that ten thousand hours won't make everyone a Grandmaster, but that every Grandmaster has spent ten thousand hours overlooks the high degree of variance among Grandmasters and particularly among young aspiring GMs.

I have spent many years training the top chess prospects in the United States as part of the Kasparov Chess Foundation's activities, which mostly focus on promoting chess in schools. Our "Young Stars—Team USA" program, cosponsored by Rex Sinquefield and his Chess Club and Scholastic Center of Saint Louis, has helped produce many junior world champions in age groups from eight to twenty, as well as several Grandmasters. One of the reasons we have been so successful is our ability to recognize talent early, sometimes even before the player has received formal training.

Competitive results are a standard indicator and are relatively easy to spot. For example, a nine-year-old with an expert rating of 2100 is far more impressive than a twelve-year-old with the same rating. Kasparov Chess Foundation president Michael Khodarkovsky, a Soviet chess coach who immigrated to the United States in 1992, has attempted to replicate aspects of the Soviet chess conveyor belt and the Botvinnik School of which I was a graduate and, later, a visiting coach. Before the program was initiated, few American kids so young were getting serious training or playing in strong tournaments frequently. Today, we can proudly say that the American juniors are one of the strongest squads in the world.

The rating outliers, to use Gladwell's term, are the kids who are performing at a level two or even three years ahead of their peers. If a twelve-year-old is performing at around 2300 that's very good, but if there's a nine-year-old with that rating, he or she is something special.

Sometimes they level off, but usually if they have outstripped their peers by several hundred points for a few years, there won't be a regression until they are faced with a career choice: professional chess and full-time training, or college.

Usually the youngsters who have demonstrated exceptional results have benefited from some combination of a strong school chess program, dedicated parents, frequent competition, and professional-grade training tools like databases. But this isn't always the case, and even these staples of any sports prodigy cannot account for the rare kids who stand out so far above their peers. One member of our program, Awonder Liang of Wisconsin, was nine years old when he first defeated a Grandmaster. At age thirteen, he was already the fifth-highest-rated player in the United States under the age of twenty-one. The next player his age to appear on the list was at number forty-nine, over 200 rating points lower. Number one on the US junior list is Jeffery Xiong, who just won the world under-twenty championship while only fifteen and is already among the overall top one hundred players in the world.

We also have more ways to measure talent than results and rating points. Before kids are accepted into the program and while they are participants, I carefully review a selection of their games. And while I won't claim a perfect record of picking winners, it is obvious to me when a young player displays flashes of brilliance. By brilliance I mean the sort of inspiration and creativity that cannot be produced by ten million hours of practice, let alone ten thousand, and often these kids have only been playing for two or three years. The talent that Gladwell admits will never permit him to become a Grandmaster at any age is distinctly present in a child of seven. What to call a child with such abilities if not "a natural in a cognitive field"?

These rare gifts don't guarantee a bright future in chess, of course. Other aspects of the game may prove too challenging. He may decide to drop the game entirely the next year in favor of soccer or Pokémon, or her parents may decide chess is a waste of time or that it's too inconvenient and expensive to travel to tournaments. But the ability was there because I saw it. I saw it on the chessboard with my own eyes that there was something very special, somewhere deep inside a few pounds of soft gray matter.

If everyone played chess, we would have a better idea of just how rare a talent for it really is. Had I been born in a place where chess was not a national pastime, would I still have a gift for a game I never learned to play, like a tree falling in a deserted forest? Might I have become a shogi champion had I been born in Japan, a rival to my colleague, the shogi legend Yoshiharu Habu? Or a xiangqi player in China or an oware player in Ghana? Or, as it seems to me, does chess require a special mix that matched my mind almost perfectly?

Despite not knowing all the rules, before I was six I solved a chess puzzle in the newspaper that had been frustrating my parents. My father, Kim, hastily got out the chess set the next day to show me how the game was played, but it always felt like I learned chess the way an infant acquires its native language. There is no luck in chess, but clearly I was lucky in my choice of birthplace and parents. My father taught me the rules before he passed away when I was seven, but he wasn't that interested in the game. It was my mother, Klara, who had been considered something of a chess ace in her childhood, though such diversions were soon pushed aside by World War II.

Lastly on talent, don't tell me that hard work can be more important than talent. This is a handy platitude for motivating our kids to study or practice piano, but as I wrote ten years ago in *How Life Imitates Chess*, hard work *is* a talent. The ability to push yourself, to keep working, practicing, studying more than others is itself a talent. If anyone could do it, everyone would. As with any talent, it must be cultivated to blossom. It can be convenient to frame work ethic as a moral matter, and certainly there is the usual intertwining of nature and nurture involved. And I would hate to provide anyone with a genetic excuse for taking it easy. But to me it has always sounded a little absurd to say that "player X has more talent but player Y wins because she works harder." Reaching peak human performance requires maximizing every aspect of our abilities whenever we can, including preparation and training, not only while at the chessboard or in the boardroom.

IN KEEPING WITH my optimistic nature, I have decided that it was good fortune, not ill, that put me in the position of being the world chess champion when computer chess came of age at last. The eighteen years I spent battling each new and better generation of chess machines added a great deal of interest to my chess career. It put me in contact with a different world of science and computers that I otherwise would never have experienced.

Of course, it was much more pleasant when I was winning these battles than when I wasn't. But I didn't have much time to contemplate this turning of the tide. The evolutionary processes that produced the human mind and the best Soviet training techniques were no match for the relentless march of Moore's law.

My first public event against computers was that 32–0 rout in a simultaneous exhibition in Hamburg. My last was in 2003 in New York City, a drawn six-game match against a PC program called X3D Fritz, in which I wore a pair of 3-D glasses and made my moves on a floating virtual reality board. Between those historical bookends, I played dozens of games against machines, some in casual exhibitions and others in serious tournaments and matches. Looking over these games now and seeing how dramatically the machines improved is like watching a child grow up in fast-forward.

I wasn't the only Grandmaster playing against computers. Beginning in the late 1980s it was the fashion to have a computer participant in tournaments, if not yet in strong Grandmaster events. In open tournaments, where anyone can play (as opposed to invitational, or "closed," events) computers started going from a curiosity to a threat. Most of these events allowed players to opt out of being paired against a computer opponent, and many did. Others, especially strong players with experience with computers, were happy to take them on. Some had more success than others thanks to a short-lived specialty called "anti-computer chess."

Every strong human player has a style, as well as different strengths and weaknesses. Understanding these things in yourself is a key component to improving as an elite player. Understanding them in your opponents is also very important, as Emanuel Lasker and his psychological insight demonstrated. Lasker understood the preferences and tendencies of his rivals better than they understood themselves, and he exploited this knowledge ruthlessly by shifting the battle into positions where he knew his opponent was uncomfortable.

Chess computers don't have psychological faults, but they do have very distinct strengths and weaknesses, far more distinct than any equivalently strong human player would have. Today, they are so strong that most of their vulnerabilities have been steamrolled into irrelevancy by the sheer speed and depth of brute force search. They cannot play strategically, but they are too accurate tactically for a human to exploit those subtle weaknesses decisively. A tennis player with a 250-m.p.h. serve doesn't have to worry very much about having a weak backhand.

That was far from the case back in 1985. Tactical calculations were still a computer strength, but only shallow sequences three or four moves deep. This was more than enough to beat most amateurs consistently, although strong players became adept at setting tactical traps that were too deep for the computers to see. It seemed paradoxical that the machine's strength of flawless calculation was also a major weakness. The brute force "exhaustive search" method of checking every one of millions of positions also meant that the search tree couldn't reach very deep. If you could find a tactical threat that struck the decisive blow four moves (eight ply) away when the computer could only see three moves (six ply) deep, it wouldn't see it coming until it was too late. We call this the "horizon effect," exploiting that the machine can't see beyond its search "horizon."

Strong humans who were aware of these machine handicaps would set up their pieces behind their pawns when playing computers, avoiding exchanges and minimizing tactical complexity as much as possible. They would prepare all their forces behind the lines, with any breakthrough far enough away that it would remain beyond the computer's horizon. The computers were strong enough not to blunder in these circumstances, but they would shuffle around harmlessly, oblivious to the mounting danger while the human player wound up for a knockout blow. Lasker would be proud.

That would never work against a decent human player. We can glance at the board and think, "I don't see any immediate danger, but my opponent is clearly massing for a big attack over there, so I should do something." We can think in generalities like "my king is weak" or "his knight is in a threatening position" and begin our move analysis from those evaluations without having to calculate everything move by move. If a brute force algorithm cannot reach deep enough to see a position in its search tree, it doesn't exist.

Another anti-computer strategy from the good old days took this horizon plan to the extreme, playing very passively and solidly until the computer created weaknesses in its own position. Having no concept of biding its time, machines would advance pawns, put pieces out of position, and generally wander without a plan unless there were concrete targets to attack or defend.

Later, programming techniques were developed that allowed programs to "fantasize" a little by looking at hypothetical positions away from the search tree, but this came at the cost of slowing the main search. Much more success was had with ways to make the search smarter and deeper with techniques like "quiescence search" and "singular extensions" that tell the algorithm to deeply examine variations that meet special conditions, such as piece captures or the king being in check. It's a slight wave toward the old Type B programs and the dream of playing chess like a human and prioritizing certain moves early on, but it's still search, not knowledge. These clever techniques went a long way toward eliminating the horizon effect in practical play, as did ever-faster chips.

Looking at the games of the best machine chess players of the 1980s now, I can say they did not play good chess. But they were increasingly dangerous because humans make so many mistakes of the kind that computers are perfectly designed to exploit. In purely chess terms, a human versus machine game is asymmetrical warfare. Computers are very good at sharp tactics in complex positions while that is a human's greatest weakness. Humans are very good at planning and what we call "positional play," the strategic and structural considerations and quiet maneuvering. This fire and ice battle is one reason these clashes were always intriguing. But at the end of the day, it's impossible to eliminate tactics against a strong opponent forever.

Again and again the pattern repeats in these matchups in which the human loses. The master, playing from years of opening knowledge and experience, steadily builds up an overwhelming position while the computer can't find a plan. Often the human player sacrifices a pawn to gain a dominant position in exchange for the material deficit. The human eventually has to find a way to cash in on his advantages to win material or attack the machine's king. As soon as that happens, as often as not, the computer finds some dazzling tactical blows and defends like a demon to reach a draw or even to win the game.

The only game David Levy lost to Chess 4.7 in their 1978 match is a good example of this demoralizing formula. Levy played a very sharp opening with black in the fourth game of the match, something that would be tantamount to suicide against a top program today. But he came out in excellent shape and, after sacrificing a pawn for a strong attack, looked set to score his third win in a row and the match. He would have to wait a few hours to collect his money, however, as he failed to find the knockout punch and the program found several tricky "only" moves—what we call it when there is only one move to avoid immediate disaster. Chess 4.7 fended off the attack and eventually even won, the first machine victory over an International Master in a serious game. To be fair to the program, it had had a completely winning position in the first game of the match before letting Levy off the hook to draw, an amusing reversal of the traditional human and machine roles.

By 1983, Thompson and Condon's Belle was the first to achieve a master rating. In 1988, HiTech, a specialized hardware machine like Belle before it and Deep Blue after it, raised the bar again by beating a strong International Master in the Pennsylvania State Championship. Harvard University started a series of human versus machine events that pitted a team of American Grandmasters against some of the top programs. The scores over the six years of the event tell the tale. In the first two, all of the humans finished ahead of all of the computers. That would not be the case in the subsequent events, although the Grandmasters still had a sizable advantage over the PC programs they were battling. Still, it was clear the computers were making steady progress. The humans won 13.5–2.5 in 1989, 18–7 in 1992, and 23.5–12.5 in the last event in 1995. They were probably wise to stop then.

In September 1988, HiTech beat the venerable American Grandmaster Arnold Denker in a four-game match, although this was the sort of victory that was all too easy to explain away. Denker was seventy-four and largely inactive and HiTech had already beaten several players considerably stronger. Denker blundered badly numerous times, lost one game in thirteen moves, and was totally lost in another by move nine. This level of play did allow the machine to display the fearsome tactical abilities for which they were becoming notorious. But if machines wanted full credit for defeating a human of the highest title, they would have to aim higher.

HiTech creator Hans Berliner's commentary after the Denker match was a preview of the sort of arrogance that many in the chess community found more than a little grating. It is entirely natural to be proud of your creation's successes, of course; possibly no less with a machine than with a child. That said, when your machine is competing against a human who has dedicated himself to this sport all his life and achieved tremendous success, one should probably keep boasting to a minimum. Berliner, rare among programmers for being a strong chess player himself, lavished praise on nearly every one of HiTech's moves in his annotations of the fourth match game against Denker. "HiTech played truly brilliantly," he wrote in *AI Magazine*, and in the game notes he scattered everywhere the exclamation points we use to indicate moves of special quality and attractiveness. All this for a lopsided game that was essentially over before the tenth move.

I will try to be a little sympathetic because in 1988 this was a fine accomplishment for a machine player, but beating an opponent who played as badly as Denker did in that game should engender modesty, not hubris. And targeting an elderly player with no experience against machines might also be seen as less than sporting. I suspect that Berliner was already becoming defensive about the more impressive progress of HiTech's stablemate at Carnegie Mellon, the grad student project Deep Thought. With a few notable exceptions, I found the chess programmers to be gracious and respectful toward their human opponents. Those who were not often appeared to have been caught up in putting the competition too far ahead of the science, or confusing their machine's chess abilities with their own.

For Grandmasters, computers were aliens among us, visiting our world at our invitation. Some of us were hostile toward them, but mostly we were curious and, occasionally, fairly compensated for these exhibitions, as Jesse Owens had been for racing against horses and cars, but it was always an awkward dance.

The great AI pioneer Donald Michie, who worked at Bletchley Park with Alan Turing cracking the Enigma code during World War II, wrote wisely about this in 1989, predicting that there could be a "Grandmaster backlash" against machine participation in tournaments:

> Chess is a culture shared among colleagues who form a human community, however adversarial the game may be in itself. After play, opponents commonly analyze the fine points together, and many find in the tournament room the mainstream of their social life. Robot intruders contribute only brute force, not interesting chess ideas. . . .
>
> Rather as a tennis professional facing a robot player able to impart spins which could never come from a human-held racket, Grandmasters will find in such opposition only obscurity. What has this to do with the skill to which they have devoted their life?

Michie also compared playing against a computer to a professional opera singer performing a "duet with a synthesizer," an analogy I appreciate very much. The love of chess, love for its art and emotion, runs deep in every Grandmaster. As I have tried to impart, the game has roots on a cultural and personal level. Being crushed by a robot that experiences no satisfaction, no fear, no interest at all is difficult to process.

And how are we supposed to feel about the bystander at the battle, the programmers and engineers, however clever they and their creations may be? They would often express satisfaction or dismay, but it was always a strange ritual. As Michie mentioned, it was odd to have no one to talk to about the game afterward, win or lose. Instead, we might huddle around the screen to see what the computer had been thinking during the game. It was hard to not recall the retort attributed to Bobby Fischer when an eager fan pressed him after a difficult win. "Nice game, Bobby!" Fischer answered, "How would you know?"

As was inevitable, the machines finally struck real gold in 1988 in California, appropriately enough. At a strong open tournament in Long Beach, Deep Thought scored the first machine tournament win against a Grandmaster, Bent Larsen of Denmark, a former candidate for the world championship. The "Great Dane" was past his peak at fifty-three, but still very strong and the loss was not the result of a terrible blunder. Not only did the Carnegie Mellon machine beat a GM, and an eminent one at that, but it tied for first in the tournament with another very strong Grandmaster, England's Tony Miles. The next year, Deep Thought crushed David Levy 4–0, as if to avenge its many fallen silicon comrades. It was 1989 and the machines had finally arrived. It was time for me to enter the arena.

CHAPTER 6

INTO THE ARENA

PEOPLE UNDERSTAND that computers are very good at calculation, and since nonplayers generally assume that chess is mostly calculation for humans as well, they were often surprised that humans could compete with chess machines at all. This was quite a flip-flop from the 1950s, when the idea of a machine playing chess sounded like science fiction. The difference in public perception was mostly due to Apple, IBM, Commodore, and Microsoft, the companies that put a computer in every home, office, and school. Computers became familiar objects with amazing powers; certainly an ancient board game should be no challenge for them.

These misperceptions, combined with centuries of romanticizing chess as an intellectual paragon, contributed to the luster of the human world champion doing battle with the machines. Chess wasn't exactly front-page news in the West, although it was accorded reasonable treatment as a sport in most of Europe instead of being relegated to the comics and puzzle pages as it often was in the United States. The tie-in between chess and the computer revolution proved very attractive for advertisers, the media, and the general public. This was no small thing for a sport like chess, which had often struggled to find sponsorship.

This was an issue even for the battles for the chess crown, although things were beginning to improve. I played five consecutive world championship matches against Anatoly Karpov from 1984 to 1990, an unprecedented series of contests that elevated the game and the attention it received nearly to the levels of the Bobby Fischer–Boris Spassky match in 1972. That match was unique, garnering far more interest and money than those of the decade before it and after it combined. It was a Cold War showdown, the brash American against the Soviet apparatus, played on the world stage in Reykjavik for hundreds of thousands of dollars, an incredible amount at the time, instead of between two Soviets in a Moscow theater for peanuts, pride, and privileges.

My first match with Karpov began in September 1984, the "marathon match" that dragged on for five months and forty-eight games before being cancelled by the World Chess Federation (FIDE) after I had narrowed the gap with two consecutive wins. When I finally took the title from Karpov in a new match in 1985, I was twenty-two, Western leaning, and eager to explore my newfound political and economic advantages as world champion. My ascent to the top of chess Olympus also coincided with the rise of Mikhail Gorbachev to the leadership of the Soviet Union, and his policies of glasnost and perestroika (openness and reform). I exploited the new environment to ask questions. If I won a tournament in France, why should I have to give most of my winnings to the Soviet Sports Committee? Why couldn't I sign lucrative sponsorship deals with foreign companies the way any other sports star in the world could? Why, I asked, in the pages of *Playboy,* no less, shouldn't I drive around Baku in the Mercedes I had won fair and square in a tournament in Germany? I led this fight not only for myself, but for other leading Soviet athletes as well. I occasionally got in trouble for voicing these "unpatriotic" opinions, but by the late 1980s the Soviet leadership had bigger problems on their hands than a renegade chess champion. And if I wasn't as reliable as Karpov, at least I was continuing his winning ways.

For our 1986 rematch, we pushed out into the brave new world and split the twenty-four-game match between London and Leningrad (now St. Petersburg). It was the first time a world championship between two Soviets took place outside of the USSR. We stood on stage with Margaret Thatcher at the opening ceremony and gave interviews in English, if usually under the watchful eye of our KGB minders. The fourth "K-K" match, in 1987, took place entirely in Seville, Spain, and I barely held on to my title by winning the last game. By the time of our fifth and final match, in 1990, it was split between New York City and Lyon, France. The Berlin Wall had fallen, the USSR wasn't far behind, and a whole new world of challenges and opportunities was opening up for me and for chess. Machines would become an exciting part of this new era.

Right around when Deep Thought became the first chess machine to become a real threat to Grandmasters at the end of the 1980s, artificial intelligence was experiencing a broad resurgence in the scientific and business worlds. The so-called AI winter that had descended after years of overpromising and subsequent disillusionment was lifting. The crisis for AI stemmed from the evaporation of the confidence so many experts in the 1970s had had in quickly discovering the secrets of cognition. Research projects and commercial AI ventures were shuttered throughout the 1980s and the AI movement had splintered. Basic science was out, practical systems were in. Understanding human intelligence was passé, getting results in a narrow field was the fashion. The maxim became "Don't make it think, just make it work."

Speaking at an AI conference in Seattle in 2001, Microsoft chairman Bill Gates reminisced about the great expectations that were in the air about artificial intelligence in the 1970s. "Microsoft was founded about twenty-five years ago, and I can remember at the time thinking, 'Well, if I go out and do this really commercial stuff, I'm going to miss these big advances in AI that will be coming very soon.' [Laughter] And so I come from the school of AI optimist. You know, I can remember being at Harvard and back then AI was the Greenblatt Chess Program and Maxima and Eliza and people literally felt that within five to ten years that some of these tough problems would be solved."

To be fair, those AI pioneers took aim at the biggest targets, like using natural language, self-teaching machines, and understanding abstract concepts. Still, their optimism sounds wildly out of proportion in hindsight. The 1956 Dartmouth Summer Research Project that launched the field of AI boldly proclaimed that great progress would be made on all of these things "if a carefully selected group of scientists work on it together for a summer." A summer!

I won't criticize anyone for dreaming big, however; it's how technology changes the world—and it doesn't happen on a fixed schedule. With a healthy kick in the pants from Sputnik, the American science and engineering community in the 1950s and 1960s was building the foundation of nearly every digital technology we depend on today, from the Internet to semiconductors to GPS satellites. If true AI turned out to be too hard a problem to solve, many other ambitious projects of the time had more success.

The story of the predecessor of the Internet, ARPANET, is an invaluable one, but it's too long and too far afield to tell in full here so I'll limit myself to one personal anecdote. In 2010, I was in Israel as a guest speaker at the Dan David Prize ceremony in Tel Aviv. Every year, the Dan David Foundation and Tel Aviv University give out prizes that "recognize and encourage innovative and interdisciplinary research that cuts across traditional boundaries and paradigms." Leonard Kleinrock of UCLA was there to receive in the category of "The Future—Computers and Telecommunications." As a slideshow presented the audience with a summary of Kleinrock's achievements, I excitedly whispered to my wife, Dasha, "That's him! That's the guy who sent the 'l' and the 'o'!"

On October 29, 1969, Leonard Kleinrock's lab sent the very first letters over ARPANET from his computer at UCLA to another machine at Stanford. They attempted to send the word "login" but the system crashed after the first two letters had gone through. A month later, a permanent link between the machines was in place. A few weeks after that, two more computers had been added, in Santa Barbara and Salt Lake City. I was familiar with the basic facts of the story and had used the ARPANET story to rebut audiences who wanted to claim the Internet wholly for the 1990s. Being able to meet the man himself was an unexpected honor.

Kleinrock, who received the 2007 National Medal of Science in the United States, developed the mathematical background for packet switching, the most elemental network building block of the Internet. His theoretical work on routing network traffic is what today's World Wide Web operates on. He points out that while it took considerable time to build the hardware and software required for the early networks, the ambition of the people working on the project was always global in scope despite the primitive nature of their early inventions. Beyond global, in fact.

On April 23, 1963, Joseph Licklider, a director at the Advanced Research Projects Agency (ARPA), sent out an eight-page memo to his colleagues, broadly describing the goals for their new project to get computers to talk to one another, and addressed it to "Members and Affiliates of the Intergalactic Computer Network." Talk about ambition! That document, and several others that followed, established the scope of ARPA's quest, including descriptions of transferring files, email, and even the potential for digital voice transmission that we would now recognize as Skype.

The Internet did not become a transformative technology, essential to many in their daily lives and economically impactful on a global scale, until over twenty years after Kleinrock sent those first letters. Email predates the Internet and was already widely used in the scientific community and on university campuses; it is the web we think of as the world-changing invention.

ARPA was founded in February 1958 by the Eisenhower administration as a response to the Soviet launch of Sputnik in 1957. ARPA's stated goal was to prevent further such surprises, a mandate that was soon expanded to the creation of similar technological advances with which to surprise America's enemies. Ironically, the vague description that was intended to help the new agency through the budget and Pentagon approval processes turned out to be ideal for funding experimental research. The military didn't want a new bunch of eggheads taking over crucial military tech sectors like missile systems, so many of the early ARPA projects went off in unexpected directions without direct military application.

Artificial intelligence was one of these directions, although progress was far slower than hoped for. In 1972, the agency acquired the "D" for "Defense" and changed its name to DARPA. Then the 1973 Mansfield Amendment limited DARPA appropriations to projects with direct military application, a heavy blow to government support of basic research in the sciences and a death blow to relatively unproductive fields like AI was turning out to be, at least in the eyes of the Defense Department. They wanted expert systems for recognizing bomb targets, not machines that could talk.

Leonard Kleinrock was still at UCLA, but he turned out to be our neighbor on the Upper West Side of Manhattan. He was gracious enough to share with me some of his thoughts on why and how ARPA (as he always insisted on calling it) fell from grace as an engine for AI and other tech innovation. His first conclusion was not surprising: the growing government bureaucracy stifled communication and innovation. "It got big," he told me over lunch. "For a while, when we had retreats you would have physicists and computer guys swapping stories and ideas with microbiologists and psychologists. Everybody could fit into one room. As it grew, that became impossible, and the different groups had little contact with each other."

Instead of a small club of brilliant (and well-funded) scientists sharing ideas in relative freedom, DARPA became an unwieldy hierarchy. This is why I chose interdisciplinary research as my area of emphasis when I joined the Oxford Martin School as a senior visiting fellow in 2013. Great new things come from cross-pollination.

Kleinrock also pointed out that the shift to military applications meant the dozens of graduate students who helped with his DARPA-funded projects were kicked out due a lack of security clearances. Pushing so many bright young minds out of important research was unacceptable to Kleinrock, who stopped taking DARPA money. In 2001, Donald Rumsfeld took over the Department of Defense with every intention of shaking things up from top to bottom. His stated desire to return DARPA to its lean, ambitiously experimental roots was mostly thwarted by 9/11 and the immediate focus of all resources on meeting the terrorist threat. DARPA turned to projects dedicated to information gathering and analysis, most infamously resulting in a public debacle over the Orwellian-named Total Information Awareness program in 2002.

DARPA never completely gave up on AI, and even had room in its budget for a little chess. If you check the fine print of the scholarly papers on Hans Berliner's machine HiTech at Carnegie Mellon, you can see it was partly funded by a DARPA grant in the 1980s. More recently, DARPA has funded contests for self-driving cars and other "practical AI" tech. Using the development of chess machines as a model, DARPA has proposed tournament competitions to develop autonomous network defense. In true Darwinian fashion, focusing on competition over basic research was bad for true AI, but very good at producing better and better chess machines. And the military always has a keen interest in intelligence analysis algorithms and smarter combat tech, which I'll return to later.

THE GRAND PREDICTIONS by the AI researchers of the 1950s and 1960s echoed those of the computer chess crowd of the same era; indeed, some of them were the very same voices. But unlike the AI researchers, chess discovered a golden ticket, the alpha-beta search algorithm that guaranteed steady improvement. Whether this was a blessing or a curse, it was tangible progress. For those investigating general purpose intelligence and other ambitious goals, there was no such concrete incremental growth of the sort needed to guarantee more graduate study programs, corporate investment, and government research grants. AI wouldn't see its spring until a movement arose that, again similar to machine chess, gave up on grandiose dreams of imitating human cognition. The field was machine learning, which had been around for years without showing very good results. What made the difference in the 1980s was data—lots and lots of data.

Donald Michie was a machine-learning pioneer himself, applying it to tic-tac-toe in 1960. The basic concept is that you don't give the machine a bunch of rules to follow, the way you might try to learn a second language by memorizing grammar and conjugation rules. Instead of telling it a process, you provide it with lots of examples of that process and let the machine figure out the rules, so to speak.

Language translation is again a good illustration. Google Translate is powered by machine learning, and it knows hardly anything about the rules of the dozens of languages it works with. Google doesn't even worry very much about hiring people with language skills. They feed the system examples of correct translations, millions and millions of examples, so the machine can figure out what's likely to be right when it encounters something new. When Michie and others tried this in the early days, their machines were too slow and their data collection and entry systems were paltry. No one could imagine that solving such a "human" problem like language could be a matter of scale and speed. They were like the early chess programmers looking at Type A brute force programs and despairing they would ever be fast enough to play competently. As one Google Translate engineer put it, "When you go from 10,000 training examples to 10 billion training examples, it all starts to work. Data trumps everything."

When Michie and a few colleagues wrote an experimental data-based machine-learning chess program in the early 1980s, it had an amusing result. They fed hundreds of thousands of positions from Grandmaster games into the machine, hoping it would be able to figure out what worked and what did not. At first it seemed to work. Its evaluation of positions was more accurate than conventional programs. The problem came when they let it actually play a game of chess. The program developed its pieces, launched an attack, and immediately sacrificed its queen! It lost in just a few moves, having given up the queen for next to nothing. Why did it do it? Well, when a Grandmaster sacrifices his queen it's nearly always a brilliant and decisive blow. To the machine, educated on a diet of GM games, giving up its queen was clearly the key to success!

This was disappointing and good for a laugh, but imagine the potential problems in the real world with machines building their own rules from examples. Turning to science fiction is often helpful, and the genre is full of accurate and insightful predictions in many fields. I hope you don't mind if I prefer to skip the killer robots and super-intelligent machine overlords of the *Terminator* and *Matrix* series. These nightmare scenarios make for good movies and good headlines, but such dystopian futures are so distant and so unlikely that talking about them distracts us from more immediate and more likely challenges. And maybe I've just had enough of battling real machines.

The 1984 movie *Starman* brings a naïve alien explorer to Earth in the form of Jeff Bridges. He's trying to blend in and learn by watching the humans around him, an extraterrestrial version of general purpose machine learning. He still makes amusing mistakes, naturally, but a more serious one comes when he takes a turn driving the car of a woman who has befriended him. Starman speeds through an intersection, causing a crash behind him, and the woman, Jenny, screams at him, leading to this exchange:

> STARMAN: Okay?
> JENNY: Okay? Are you crazy? You almost got us killed! You said you watched me, you said you knew the rules!
> STARMAN: I do know the rules.
> JENNY: Oh, for your information, pal, that was a yellow light back there!
> STARMAN: I watched you very carefully. Red light stop, green light go, yellow light go very fast.
> JENNY: You'd better let me drive.

Perfect. Like a chess program trained to imitate Grandmaster brilliancies that gives away its queen, learning the rules by only observation can lead to catastrophe. Computers, like visiting aliens, don't have common sense or any context that they aren't told or cannot build. Starman was not wrong, exactly; he just didn't have enough data to figure out that accelerating at a yellow light requires much more context. Even the petabytes of data used by Watson and the billions of examples that pour into the bottomless maw of Google Translate often lead to strange results. As is usually the case in science, what goes wrong teaches us more than what goes right.

Watson's *Jeopardy* response about a 1904 Olympic gymnast with "this anatomical oddity" was revealing. Human champion Ken Jennings buzzed first and, clearly unsure, guessed "only one hand" and was wrong. Watson then answered simply "leg" (actually, "What is leg?" in the show's vernacular), with that response a heavy 61 percent favorite in its evaluation. It was very clear what had happened. Gymnast George Eyser was missing a leg, undoubtedly his claim to fame. Watson's search therefore turned up lots of results with Eyser's name and the anatomical word "leg." So far, so good. But the machine got wrong what Jennings got right because it couldn't understand that having a leg is not an oddity. Jennings was wrong in a human way, with a logical assumption that lacked data. Watson was wrong in a machine way, having the right data but none of the broad context that functions as common sense in the human mind.

I don't know if Watson was programmed to pay attention to any of the answers that the humans gave before it, but if it had been, perhaps it could have figured things out by combining its correct data with Jennings's correct assumption. Definitely the third player, another human champion, should have done this. Perhaps since it was the first show, he wasn't yet confident in Watson's accuracy. Had he done so it would have been an excellent demonstration of my ideas on how humans and artificially intelligent machines can work together.

Anyone who travels as often as I do knows how difficult accurate translation is. Long before intelligent machines were available to butcher languages for us, signs and menus around the world were full of bizarre phrases likely assembled directly from bilingual dictionaries. A "lounge for the weak" at an airport, a "plate of little stupids" at a restaurant. Now Google and other services will translate entire webpages on the fly, usually with enough accuracy to get the gist of a news story in just about any major language.

THERE ARE MANY GLITCHES, of course. My favorite is чят, a purposely distorted Russian slang word for online chat (also pronounced *chat*), casually used on social media to refer to one's audience, the way people say "Hello, tweeps" on Twitter. But somewhere deep down inside of Google Translate's Russian database, these three Cyrillic letters have been associated with something very different. I discovered this when viewing my Twitter feed translated automatically on a friend's computer, where Russians were saying, "Hello, sensitive nuclear technologies"! Using Google again you can find some obscure government papers in which ЧЯТ is indeed used as an acronym for чувствительных ядерных технологий, or "sensitive nuclear technologies."

This is unlikely to cause a panic because the humans seeing it likely have enough common sense of their own to know something strange is going on and to blame the machine translation instead of raising the nuclear alert level to DEFCON 2. But what if military AI algorithms are making that decision, not humans? What about the security agencies that rely on computer acquisition and analysis of terrorist "chatter"? They aren't going to show each tweet to a human to double-check; that would be too slow to be useful. Instead, they might raise a flag because a bunch of Russians are talking about nuclear technology on social media.

New tech terms and slang are always going to be very hard for machines to figure out, and, like a chess machine or a trivia robot, they have no way to apply practical chances or common sense. They must simulate it. There is only an evaluation, a number representing a confidence factor. A machine learning system is only as good as its data, the same way a chess program's opening book is only as good as the games fed into it. Errors are reduced by quantity leading to quality, keeping the good examples and discarding the bad a billion times per second, although there will always be anomalies and, of course, sensitive nuclear technologies!

Machine learning rescued AI from insignificance because it worked and because it was profitable. IBM, Google, and many others used it to create products that got useful results. But was it AI? Did that matter? AI theorists who wanted to understand and even replicate how the human mind worked were disappointed yet again. Douglas Hofstadter, the cognitive scientist who wrote the hugely influential book *Gödel, Escher, Bach: An Eternal Golden Braid* in 1979, has stayed true to his quest to comprehend human cognition. Consequently, he and his work have been marginalized within AI by the demand for immediate results, sellable products, and more and more data.

Hofstadter expressed his frustrations in a great 2013 article about him by James Somers in the *Atlantic*. Hofstadter wanted to ask, why conquer a task if there's no insight to be had from the victory? "Okay," he says, "Deep Blue plays very good chess—so what? Does that tell you something about how we play chess? No. Does it tell you about how Kasparov envisions, understands a chessboard?" A brand of AI that didn't try to answer such questions, however impressive it might have been, was, in Hofstadter's mind, a diversion. He distanced himself from the field almost as soon as he became a part of it. "To me, as a fledgling AI person," he says, "it was self-evident that I did not want to get involved in that trickery. It was obvious: I don't want to be involved in passing off some fancy program's behavior for intelligence when I know that it has nothing to do with intelligence. And I don't know why more people aren't that way."

Not to be cynical, but Google's current market cap of over $500 billion is probably one reason. Another, as several experts, including Watson's Dave Ferrucci and Google's Peter Norvig, say in the article, is that they wanted to take on problems they could solve. Human intelligence is an incredibly hard problem and machine learning works. But for how long? The law of diminishing returns is already having an impact. Getting a machine system to a 90 percent effectiveness rate may be enough to make it useful, but it's often even harder to get it from 90 percent to 95 percent, let alone to the 99.99 percent you would want before trusting it to translate a love letter or drive your kids to school.

The machine-learning approach might have eventually worked with chess, and some attempts have been made. Google's AlphaGo uses these techniques extensively with a database of around thirty million moves. As predicted, rules and brute force alone weren't enough to beat the top Go players. But by 1989, Deep Thought had made it quite clear that such experimental techniques weren't necessary to be good enough at chess to challenge the world's best players. What was necessary was speed and more speed, and the custom chips designed by Feng-hsiung Hsu at Carnegie Mellon were delivering it. After it beat Bent Larsen, and also Tony Miles in an exhibition game, I felt that it could be an interesting new challenge, and so challenge it I did.

MY TWO-GAME MATCH against Deep Thought took place in New York City on October 22, but I was the only player there in person. As is often the custom, the machine itself was hundreds of miles away, connected to the site by a relay and an operator who made the moves on a regular board and clock. The Deep Thought team had that month been hired by IBM, which would soon lead to millions of dollars in investment and technology as well as a name change to Deep Blue. And so this mini-match was sponsored by AGS Computers, a software company in New Jersey whose chairman was a chess enthusiast and who had also sponsored the HiTech match with Denker.

One of the problems with playing against computers is how quickly and how often they change. Grandmasters are used to preparing very deeply for our opponents, researching all of their latest games and looking for weaknesses. Mostly this preparation focuses on openings, the established sequences of moves that start the game and have exotic names like the Sicilian Dragon and the Queen's Indian Defense. We prepare new ideas in these openings, and look for strong new moves ("novelties") with which to surprise our opponents. This is particularly effective if you can find something nasty in one of his favorite lines, since you can reasonably expect to reach that position.

I'll go into more detail on how computers navigate these openings in the Deep Blue chapters, but I'll point out now that they rely on a database of moves derived from human play, called an "opening book." These books have evolved over the years to allow the machines more flexibility, but the basic idea is what it sounds like, a book of openings it follows more or less blindly until it "runs out of book" and has to think for itself. This is effectively similar to how I do it, relying on memory to select the opening lines I prefer until I run out of book and am on my own.

I can say without any false modesty that I was the best-prepared player in the history of chess. Even when I was very young I enjoyed studying the openings and searching for improvements to add to my arsenal. The exciting cut-and-thrust tactics of the middlegame get most of the attention, but the tenacity and ingenuity required to find a new idea in the well-trodden paths of the opening always attracted me. I studied my opponents' openings comprehensively in search of weaknesses and kept huge database files full of novelties and analysis. Even strong opponents would sometimes avoid playing their favorite openings against me, fearing a powerful novelty. When I retired from professional chess in 2005, a joke went around that I should auction off my laptop full of my precious databases.

I enjoyed hearing the urban legends about how I had a team of Grandmasters shackled in a basement producing novelties for me 24/7, when it was always just me, my trainer Yuri Dokhoian, and Alexander Shakarov, who had worked with me since 1976 and who archived and maintained these decades of precious intellectual property. I didn't enjoy it as much when critics would say disapprovingly that I had "won the game at home," when I gained an advantage from a nice piece of preparation. I accept reserving the highest praise for over-the-board brilliance, but there is nothing to be ashamed about outpreparing your opponent. Such skepticism might be a little more pertinent today, when instead of a galley of Grandmasters, every professional player prepares with the help of a super-strong engine. It is still the result of human labor, using the machine as a tool, but it's slightly hollow when a devastating new idea comes from a silicon brain instead of your own.

Having a computer opponent short-circuits much of this opening preparation. Even if you go over every game the machine has ever played, the operator can simply load an entirely new opening book, or change a few values, and the computer will play a set of openings it's never played before. And it will play them perfectly, since it has none of the human concerns about recall. They are as vulnerable to novelties as a human, however, since if a move is in its book it will play it instantly from the database, which has led to a few amusing bloopers. In one computer championship, a machine blundered a full piece early in the game but its opponent declined to capture it. Both had the same flawed line in their opening books. These days all the books they use have been thoroughly checked and improved by the engine to make sure it won't find itself in a lost position without even starting to think for itself.

If being able to access a book with gigabytes of opening moves sounds like an unfair advantage for the machine against a human, you are my kind of person. It always seemed strange to me that the computer essentially skips an entire phase of the game, never having to figure out how to develop its pieces or establish a pawn structure. The opening phase combines subtlety and creativity with long-term strategic planning, all things computers are bad at. But thanks to opening books, the computer simply skips it and goes to work in the middlegame, where its tactical prowess is at its best.

Unfortunately, there is no fair alternative to opening books either, at least not without changing the rules in some way. Chess openings have been empirically developed over decades and are studied and memorized. Even a weak tournament player can remember enough opening moves to reach a playable position without having to do any real thinking. (This is a bad habit that I criticize as a coach, since it leaves the player without any real understanding of the position once he's out of book.) The openings are a huge part of chess and simply removing them from the computer would be an unfair advantage for the humans. It would also produce very foreign-looking games, since the machines would tend to play the same straightforward developing moves every time if left to their own devices. This is easy to test by just turning off the opening book in your favorite chess program. Today's programs are still nearly impossible to beat, but it gives a reasonable chance for a strong human player to control the early flow of the game when the book is off.

The openings aren't the only thing that can change in a computer opponent from one game to the next. It's easy to tweak a few values to make the program more aggressive, for example. There could be six different machine "personalities" stored away so you never really face the same opponent twice in a match of six games. Again, between two computers this is not that relevant, but experienced humans are used to profiling their opponents, and for me it was a critical part of the game.

Most of all, computers get stronger. The version of Deep Thought I played in 1989 was already significantly upgraded over the one that had beaten Larsen in Long Beach the year before. Its parallel hardware-based design meant they could keep adding more and more chess chips and computer power as soon as it was available. It had six processors and could search over two million positions per second, far more than any previous machine. These big numbers all start to sound the same after a while, so here is what the Deep Thought team wrote about the relationship between search depth and chess strength in a 1989 article:

> The ascent of the brute-force chess machines back in the late 1970s made one thing crystal clear: there is a strong causal relationship between the search speed of a chess machine and its playing strength. In fact, it appeared from machine self-test games that every time a machine searches one extra ply, its rating increases by about 200–250 rating points. Since each extra ply increases the searched tree size by five to six times, every two-fold increase in speed roughly corresponds to an 80–100 rating point gain. Ratings obtained by machines against human players indicate that this relationship holds perhaps all the way up to the Grandmaster level where Deep Thought currently resides. The presence of this causal relationship was the reason the project was started in the first place.

In other words, faster means deeper and deeper means stronger, and that was all that really mattered. You can chart the progress of chess machines with rating as the *y* axis and the number of positions searched per move as the *x*, and it makes for a nice diagonal line. Starting with Chess 3.0 in 1970 at around a 1400 level, to Chess 4.9 at 2000 in 1978, Belle breaking 2200 in 1983, HiTech at 2400 in 1987, and Deep Thought at a Grandmaster-level 2500 in 1989. The chips get smaller and faster, the search goes deeper, and the rating rises.

While the engineering was still a challenge, that dismal equation again illustrates why so many people were disillusioned with how far from its AI roots machine chess had gone. While charting that impressive rating climb in 1990, machine intelligence expert and chess master Danny Kopec lamented, "Due to the competitive priorities of most programs, little is revealed about how a program finally selects one move over another. This largely explains why computer chess has appeared to advance primarily as a competitive sport (performance driven) rather than as a science (problem driven)."

On October 22, 1989, I wasn't thinking about whether or not Deep Thought was intelligent or not, only how strong it might be. I assumed improvements had been made over the version that had beaten the strong English GM Tony Miles in an exhibition game. I had recently broken Bobby Fischer's longstanding rating record of 2785 and arrived at the board unafraid. I had been able to review the machine's previous games the day before the match, although, as I said, you could never be sure how much the machine had changed in recent months, or even days. Murray Campbell of the Deep Thought team had provided some of the games, a nice gesture in keeping with the friendly and exploratory spirit of the match. And it seemed only fair. After all, it could analyze every game I'd ever played and there was no chance I was going to upgrade my processors right before the match.

My preparation told me that it was strong, perhaps even warranting its 2500 estimated rating, which is the minimum to achieve the Grandmaster title. I would be the heavy favorite, but I estimated that in a ten-game match it would likely draw or even win a game or two. There was a lively crowd at the New York Academy of Art where the event was hosted, and I was happy to play up my role of humanity's champion for the first time. "I don't know how we can exist knowing that there exists something mentally stronger than us," I reportedly said at the opening, a statement more based on hype than logic, I would say now.

That wasn't my last rash piece of rhetoric regarding computer chess, although I would have been fine had I stopped with computers. In an interview around that time, I predicted that a computer would become world champion before a woman did, which turned out to be accurate. It was interpreted as a sexist slight, which it wasn't. There just weren't any women on the horizon who showed the potential, and that would be the case until the youngest of the three remarkable Polgár sisters from Hungary, Judit, broke into the elite a few years later, eventually reaching top-ten status.

At least I managed to back up my tough talk at the chessboard that Sunday afternoon in New York City. With the black pieces in the first game, I slowly built up a dominating position. By move twenty, I could see that I was strategically winning; it was only a matter of keeping control of the position until I could break through. The games were played at a relatively brisk pace of ninety minutes per side, quite a bit faster than the two and a half hours that was then standard for a classical game of chess. This was to the computer's advantage, since I would have less time to check my calculations, but it would be enough.

I centralized my forces and advanced my pawns against its king while Deep Thought could do little more than wait for the ax to fall. I knew that if there was any single chance to escape, the computer would find it, so I didn't rush things. A Grandmaster faced with such a pathetic and passive position would do just about anything to break free in order to at least have a chance of confusing the issue. Humans understand that risking a quick demise in exchange for a 5 percent chance of escaping is better than a 100 percent chance of a slow death with no counterplay.

Computers, on the other hand, don't understand general concepts like practical chances, however. They will always play the best move in the search tree and can do no other. Poker robots may have other ideas, but chess machines cannot bluff. One would never intentionally play an inferior move in the hope its opponent wouldn't find the refutation. A partial exception is if the programmers change its settings in advance to play for a win at all costs, telling it that a draw must be avoided. This is called the "contempt factor" setting, and can encourage the machine to play riskier continuations instead of settling if the position is a draw. Essentially it makes the computer super-optimistic about its own position or, as the name implies, contemptuous of its opponent's abilities.

Deep Thought didn't have much of a chance to be optimistic or contemptuous in our first game and, despite its typically dour defense, I eventually smashed through to win in fifty-two moves. I'm a little chagrined now to see that I did not play the best moves throughout despite my large advantage, and at one point Deep Thought could have put up a much stronger defense. After the game I bragged that "a human who got beaten like that wouldn't come back for more," but of course the machine could not be intimidated and soon I sat down with the white pieces for the second game.

White moves first in chess and, at least at the expert level, this confers an advantage similar to that of serving in tennis. White wins about twice as often as black at the professional level, although half of all games finish drawn. White can usually define the battleground and I used it to offer Deep Thought a "poisoned pawn" in the opening, a tempting offer of material that computers were still too eager to grab. Sure enough, the machine took the bait and was soon in great difficulties as my pieces swarmed over the board. My attack on its king forced it to give up its queen on move seventeen and after that it was a mop-up operation. Any human would have resigned against me in good conscience at that point, but machines don't have to worry about that. Their operators usually figure they have nothing to lose by playing on even when the machine's evaluation shows that it knows it is completely lost. Considering how tricky computers can be against humans, this isn't unreasonable, only annoying.

The operator resigned on move thirty-seven and I received a nice ovation from the very pro-human audience. My first foray into serious man-machine chess had been an easy and enjoyable success and even the local tabloids covered the match. "Red Chess King Quick Fries Deep Thought's Chips" wrote the *New York Post*, with an anachronistic Cold War jab. The Deep Thought team couldn't have been happy with the way their machine played, even if they hadn't been expecting a different result.

From reading the programmers' comments about the match now, I see that the old chess joke about never having defeated a healthy opponent has a parallel in computer chess: I've never beaten a program without a bug! Apparently, there was a glitch in the code that weakened its play, a "castling bug" they did not discover for several weeks. This, as you'll see, would become a theme. I also learned that Hsu had adjusted the machine between games to get it to play more slowly, highlighting how misguided it is to think you've learned something about a machine opponent after one game—it could play very differently just an hour later.

I honestly don't recall any particular psychological impact of playing my first serious games against a computer opponent. It was different, but not yet ominous. I think I was so confident that I did not feel the usual tension I would have against a Grandmaster. It felt more like a friendly exhibition, or a sort of science experiment. This wouldn't be the case in the coming years, however, as the machines got stronger and began to appear in serious tournaments where money and prestige were at stake, not merely the future of humanity.

CHAPTER 7

THE DEEP END

I AM A SORE LOSER.

I want to clear that up right at the start. I hate losing. I hated losing bad games and I hated losing good ones. I hated losing to weak players and I hated losing to world champions.

I have had sleepless nights after losses. I have had angry outbursts at award ceremonies after a bad defeat. I have been annoyed to discover that I missed a good move in a game I lost twenty years ago when analyzing it for this book.

I hate to lose, and not just at chess. I hate to lose at trivia games. I hate to lose at card games. (My complete lack of a poker face is why I rarely play them.)

Being a sore loser is not the attribute I'm most proud of, nor am I ashamed of it. To be the best in any competitive endeavor you have to hate losing more than you are afraid of it. The thrill of victory is wonderful, although I think any elite sportsman gets used to that feeling at a very young age. Everyone has different methods for finding motivation, especially over a long career. But no matter how much you love the game, you have to have to hate to lose if you want to stay on top. You have to care, and care deeply.

A database can bring up a list of practically every serious game of chess I've played since I was twelve years old, over twenty-four hundred games. Of those, I lost roughly 170 times. Counting only the tournament and match games over the course of my twenty-five-year professional career, starting when I was seventeen, the number of losses drops to around half that. If I was never a good loser, it was partly because I never had the chance to get good at it. In 1990, English

Grandmaster Raymond Keene wrote a book called *How to Beat Gary Kasparov* that collected all of my defeats up to that point. The book's introduction begins: "Beating Gary Kasparov at chess is considerably more difficult than climbing Mount Everest or becoming a dollar billionaire. . . . I learnt that it was six times easier to reach the peak of Everest . . . five times easier to acquire more than $1,000,000,000." Those few who defeated me might wonder if they should have gone into a different line of work.

I want to get all this out of the way because my attitude about losing inevitably comes up in any discussion of my match with the IBM supercomputer Deep Blue. To be more precise, my rematch with Deep Blue in 1997.

I am resigned to the fact that almost no one remembers I beat Deep Blue in our first match in 1996. "This Day in History" calendars don't have entries for all the failed attempts to fly across the Atlantic before Charles Lindbergh succeeded in 1927. When the 1996 match is remembered at all, it's because my loss in game one was the first time a machine had beaten the world champion in a classical time control game. Prior to that, I had played quite a few games against machines at faster time controls and lost a number of them. What we call "rapid" games allow between fifteen and thirty minutes per player for the whole game. Faster still is "blitz" chess, in which the players have five minutes or even less on the clock to begin the game. There is even "bullet" chess of just one or two minutes, which almost turns chess into an aerobic activity.

At least since the 1970s, the faster the game, the greater the advantage for a computer against a human. Grandmasters may play largely by intuition, but chess is a concrete game in the end. Without the time to calculate properly against a machine that is checking millions of positions per second, a blitz game can quickly become a bloodbath. The slight inaccuracies and tactical oversights humans make routinely against each other at fast time controls are instantly punished by the machines, and they never return the favor.

After beating Deep Thought in 1989, a few years went by before I played another machine opponent in a public match. Partly this was because there was no market for me beating up on computers when they clearly had some work to do before really challenging me, and my time was valuable. I narrowly won my fifth world championship match against Karpov in 1990 while dealing with the sudden collapse of my home country. Along with thousands of others, my family and I had been forced to flee Baku in the face of Armenian pogroms as the Soviet Union unraveled.

But I was keeping an eye on the machines' progress. I had the latest programs installed on my personal computer, using them for analysis on occasion and playing against them for fun now and then. They didn't play good chess, but programs with names like Genius and Fritz were already tactically very dangerous even on an average home PC or laptop. One moment of human inattention in a quick game and pow, it was over.

I also crossed paths with Deep Thought again, in 1991 at a computer exhibition in Hanover, Germany. The machine's team had lost and gained a few members in its transition to becoming a major IBM project. Feng-hsiung Hsu and Murray Campbell were still the team leaders and they were both in Hanover, where Deep Thought had been invited to participate in the strongest tournament yet to include a machine. It was a closed event, with six German Grandmasters and a strong International Master, with an average rating of 2514.

Now with the formidable resources of IBM behind him, Hsu was still working on his upgraded dream machine with a thousand VLSI chips, but it wasn't yet ready. Deep Thought was still the strongest machine in the world and was expected to be a contender in Hanover, based on its past performances. It was a little surprising that it finished next to last with 2.5/7, winning two, drawing one, and losing four. The team blamed two of the losses on mistakes in the opening book (another reoccurring theme), although looking at its Hanover games now, it also just didn't play very good chess.

Of more interest was a little test for me, proposed by my friend Frederic Friedel, who was one of the Hanover event's organizers. I was shown the games from the first five rounds of the tournament to see if I could figure out which player was Deep Thought. It was a chess twist on the Turing test, to see if a computer could pass for a Grandmaster. I managed to pick out two correctly and narrowed down another round to two games before choosing the wrong one, so three of the computer's five games passed the test. To me, this was a better indicator of computer chess progress than its score in the tournament. Some of its games followed the old patterns of terrible strategic play and unseemly greed balanced out by startling tactics. But other games just looked like chess, if still far from the world championship level.

I also thought this was interesting because I could imagine one day turning the tables. In ten years' time, roughly my guess for when a computer would be strong enough to beat me, would a super-strong machine be able to analyze human games insightfully? I spent a lot of time scrutinizing the tendencies and weaknesses of my opponents, but I was aware that this analysis was colored by my own tendencies and weaknesses. Machines, on the other hand, were objective. Chess engines were already proving to be useful for assisting in analysis, if mostly only for a tactical "blunder check." But once they were strong enough, I thought, maybe they would be able to detect patterns and habits in human games, both in my opponents' and my own.

This idea never really got off the ground, partly because the potential market for it was so small. There are only a few hundred players in the world who play the same opponents regularly enough to need to prepare for them specifically on a regular basis. ChessBase did eventually add some useful features like automatically building data-based player profiles, including their favorite openings and selected games. These were more time-savers than analytical tools, however. There was no advanced tendency breakdown like "often makes mistakes when his king is under attack" or "likes to trade queens when playing with black." The thought of such in-depth profiling also made some players a little uncomfortable, even though the data was all publicly available—their own games. I would love to know what a machine would say about me and my games.

I'm also very interested in what data-driven computer analysis of human behavior can do for fields like psychology, or in my realm of decision making. No one reading this would want to hand over all their texts, email, social media posts, search history, shopping history, and the rest of the long digital trail we create hourly, at least, not to a human. But different apps and services already have all that information, for better or worse, and I'm sure that enough data and enough crunching would find many fascinating correlations, perhaps even diagnose things like depression or the early signs of dementia.

Facebook has suicide prevention tools that allow friends to flag posts for staff review and possible referral, but this requires human intervention. Fitness trackers are already monitoring everything from sleep habits to heart rates to calories burned. Google, Facebook, and Amazon probably know more about you than you know about yourself already, but people would be unnerved by seeing that analysis reflected back at them, perhaps revealing uncomfortable truths.

There are countless privacy issues to be negotiated anytime such data is accessed, of course, and that trade-off will continue to be one of the main battlefields of the AI revolution. I would want to know what a machine says about my chess or my mental and physical health, but would I want anyone else to know? You might want your family and your doctor to have all this information, but what about your insurance company or your employer? Social media reviews are already part of the hiring process in some companies. Anti-discrimination laws in the United States make it illegal to ask applicants about age, gender, race, and health, but algorithmic social media analysis can identify those in a split second, as well as make very accurate guesses at things like sexual preference, political leanings, and income level.

History tells us that eventually the desire for services wins out over a vague desire for privacy. We like sharing personal information on social media. We like to have books and music recommended to us by the algorithms of Netflix and Amazon. We won't give up GPS maps and directions even though using them means dozens of private companies know where we are practically every minute of the day—information that can also be accessed by governments and courts. When Gmail introduced ads based on scanning the content of people's email there was a collective shock, but it didn't last long. It's only an algorithm and, if you're going to see an ad, wouldn't you rather see one you're interested in than one you aren't?

This is not an argument for surrendering to Big Brother. Coming from the country on which George Orwell based his novel *1984*, I am particularly sensitive to any encroachment on individual freedom. Surveillance can be an instrument of security or of repression, especially with the sophisticated tools available now. All the wonderful communication technology we depend on today is agnostic, neither good nor evil. Assuming that the Internet would magically set everyone free, as some appeared to believe, was foolish. Modern dictatorships and other political cliques are tech savvy and have learned how to limit and exploit these powerful new mediums. I'm glad privacy advocates are on the job, especially regarding the powers of the government. I just think they are fighting a losing battle because the tech will continue to improve and because the people they are trying to protect won't defend themselves. The barrage of privacy notices has become like all the disregarded warnings about the dangers of trans fats and corn syrup. We want to be healthy, but we like doughnuts more. The greatest security problem we have will always be human nature.

Technology will continue to make the benefits of sharing our data practically irresistible. Digital assistants like Amazon's Echo and Google Home listen to every word and sound in the home and people are buying them by the millions. Utility always wins. Even more invasive tech, like microsensors in our plumbing or implanted in our food or bodies, will likely be deployed first in countries with weak privacy regimes, especially in the developing world. When the results come back and show that the economic and health benefits are tremendous, the floodgates will open everywhere.

Our lives are being converted into data. This trend will accelerate as the tools become vastly more powerful and it will happen both voluntarily in exchange for services and due to the increasing public and private demand for security. This cannot be stopped, so what matters more than ever is watching the watchers. The amount of data we produce will continue to expand, and largely to our benefit, but we must monitor where it goes and how it is used. Privacy is dying, so transparency must increase.

WITH ALL the attention going to massive parallel-processing beasts with specialized hardware and custom-designed chips, there was also a PC chess revolution going on. Thanks to a growing programming community being able to share ideas on the Internet and to the ever-faster CPUs coming out from Intel and AMD, personal computers running MS-DOS and Windows were becoming very strong. By 1992, they were surpassing the strength of most of the popular stand-alone chess machines, the all-in-one chess computers built into a board made by companies like Saitek and Fidelity, with names like Mephisto and even the Kasparov Advanced Trainer.

An endorsement message from me that accompanied some models in the late 1980s said, "I wish you enjoyment and satisfaction from your Kasparov chess computer—and who knows, maybe we'll meet in combat across the chessboard in the future!" I played long enough for this to come true, and more than one young player I have faced in exhibitions has brought a Kasparov chess computer with him for me to sign.

For those too young to remember, the capabilities of personal computers in the early 1990s were never enough for what you needed to do. Even spending $5,000 on a top-of-the-line machine soon left you scrambling for more RAM, more storage, and a faster CPU. And nothing taxes processing power like a chess engine. It will happily use 100 percent of the processor, and all four, ten, or twenty of a modern CPU's cores. After fifteen minutes of running an engine my old laptop would get hot enough to double as a toaster. Even today's superpowered machines can be slowed to a crawl by a chess engine grabbing every available CPU cycle for its search.

PC programs were, and are still, far slower than the specialized hardware machines like Deep Blue, often by several factors. They compensated by being much smarter, and by using optimized programming techniques to extend the search far deeper than it could get by simple exhaustive search. They are all still Type A brute force programs, but a great deal of finesse has been added to the brute over the years. Using a multipurpose CPU allowed for more programming creativity and adaptability, and the commercial chess engines were competing constantly and tuning their evaluations, often with the help of Grandmasters. Meanwhile, although it had controller hardware that could be adjusted, Deep Thought's special chess chips were set in stone once fabricated, even though that stone was silicon.

Hardware speed depends greatly on circuit simplicity, as the Deep Thought/Deep Blue team wrote in an article about their machine in 1990. "Sacrifices in the knowledge content of the evaluation function were deemed justifiable if they simplified the circuit design significantly." They also acknowledged that "the best commercial chess programs appear to have measurably better evaluation than the research ones at this point in time." This sounds bad, but it actually gave them reason to expect greater improvements down the line when they had the opportunity to make the next generation of chips and to improve Deep Thought's evaluation function.

In 1992, I played a long casual blitz match against one of this new generation of PC programs, one that would go on to become nearly synonymous with PC chess engines. Fritz was published by ChessBase, which explains the sardonic German nickname. Its creator was a Dutchman, Frans Morsch, who had also written programs for tabletop chess machines like Mephisto. As such, he was used to having to cram tightly optimized code into very limited resources. He also helped pioneer several of the search enhancements that allowed chess machines to keep improving despite the increasing branching factor that was supposed to slow them down.

One of these is worth a brief technical detour because it's an interesting example of how machine intelligence has been augmented in ways that have nothing to do with the workings of the human mind. Called the "null move" technique, it tells the engine to "pass" for one side. That is, to evaluate a position as if one player could make two moves in a row. If the position has not improved even after moving twice, then it can be assumed that the first move is a dud and can be quickly discarded from the search tree, reducing its size and making the search more efficient. Null moves were used in some of the earliest chess programs, including the Soviet Kaissa. It's elegant and a little ironic that algorithms designed on the principle of exhaustive search are augmented by being less exhaustive.

Humans use a very different heuristic when making plans. Strategic thinking requires setting long-term goals and establishing milestones along the way, leaving aside for the moment how your opponent, or business or political rivals, might respond. I can look at a position and think, "Wouldn't it be great if I could get my bishop over there, my pawn up there, and then work my queen around to join the attack." There are no calculations involved yet, only a type of strategic wish list. Only then do I begin to work out whether it's actually possible and what my opponent might do to counter it.

Programmers who worked on human-style or "selective search" Type B chess programs had visions of teaching machines to do this sort of goal setting. Instead of only working through the tree of possible moves, the program would also look at related hypothetical positions and evaluate those. If they were good, it would raise the values of elements in those positions in its search. It improved the quality of the evaluation in many cases, but it rendered the search so slow that results suffered, the sad tale of Type B programs in general.

More success was had with another method for allowing machines to extend their thinking into the hypothetical outside of the direct search tree. Monte Carlo tree search simulates entire games played out from positions in the search and records them as wins, draws, or losses. It stores the results and uses them to decide which positions to play out next, over and over. Playing out millions of "games within the game" like this was not particularly effective or necessary for chess, but it turned out to be essential in Go and other games where accurate evaluation is very difficult for machines. The Monte Carlo method doesn't require evaluation knowledge or hand-crafted rules; it just keeps track of the numbers and moves toward the better ones.

With so many interesting ideas to improve the output of intelligent machines, you can understand why tackling things like how the human mind works and the secrets of consciousness could fall to the side. What matters most, the process or the results? It's always results that people want, whether it's in investing, security, or chess. As many of the programmers themselves lamented, this attitude was good for making strong chess machines and bad for anything to do with science and artificial intelligence. A chess machine that thinks like a human and loses to the world champion isn't going to make the news. And when a chess machine beats the world champion, nobody cares how it thinks.

And lose to a machine I finally did, to Fritz 3 at a blitz tournament in Munich in May 1994. The tournament was sponsored by Intel Europe, which had thrown its considerable weight behind the new Professional Chess Association (PCA) I had launched with my colleague and world championship challenger Nigel Short the year before. Along with many of the best players in the world, the event also included Fritz 3 running on a new Pentium chip. This was just the sort of promotion and sponsorship for chess that I had dreamed of when I saw all the publicity my 1989 match with Deep Thought received.

I had played quite a few games against Fritz's predecessor in an informal blitz match in Cologne in December 1992. Frederic Friedel says I played thirty-seven games against his beloved pet, as I poked and prodded it like a lab animal, pointing out when it made a particularly good move or chose a weak plan. It was far from the savage beast it would become, but it wasn't tame either. I lost nine times with a couple of draws, winning around thirty of the games.

Munich was another story. It was a serious tournament despite being blitz and one I fully expected to win, with or without the presence of a machine player. After a slow start, I scored eight straight wins, but Fritz 3 was right there with me, setting up our head-to-head encounter. I played aggressively in the opening and after just a dozen moves was rewarded with a crushing position. Then began the script that would become all too familiar to human players facing machines for the next decade. I played one lazy move and it counterattacked. Annoyed at my slip, I decided to sacrifice material, a rook for a bishop, in order to keep my initiative. The position was roughly equal, but in a blitz game I could not summon the accuracy to make good on my chances.

Despite a mutual blunder toward the end, where the machine and I both missed a chance for me to draw, Fritz 3 held on to win.

It was only blitz, with five minutes per side, but it was still the first victory over the world chess champion in a serious game by a machine. If not the moon landing, it was at least the launch of a small rocket. Fritz 3 and I ended up on top at the end, an impressive result for the machine. It was also an excellent silver lining, as I would get to meet it in a match for the title and exact my revenge. Here I managed to focus better and completely crush it, taking the play-off with three wins and two draws. I was even completely winning in one of the draws, but had no time on my clock to prosecute an easily winning position with a queen versus a rook.

Things did not work out so well for me a few months later when I met another PC program in an Intel PCA tournament, Chess Genius by Richard Lang. The London event was a rapid chess knockout event, with twenty-five minutes per side. I was paired against Genius in the very first round, which of course drew a lot of attention. It still wasn't a classical time control game, but the stakes were high. Whoever lost the two-game mini-match would be knocked out of the tournament, which was part of a Grand Prix series, so every point counted.

I got an excellent position with the white pieces in the first game, but missed a move that allowed the machine to equalize the position. It was then I committed another cardinal sin when playing against a computer: pushing too hard. Instead of acquiescing to the logical draw and moving on to the next game, I tried to keep the simplified position alive and immediately had cause to regret it. A surprising series of queen maneuvers by Genius left my king and knight in an awkward position and I ended up losing a pawn, and then the game. It was a brutal turn of events, and you can see my shock if you look up the clips from the game on YouTube.

Despite my blunder, I had every expectation of coming back and beating it with black in the next game and then to win the tie-break and move on in the tournament. I again got a very good position and this time won a pawn to enter another queen plus knight endgame. But Genius found a long series of improbable queen maneuvers that prevented me from advancing my pawns. Head in my hands, I had to agree to a draw. I was out. It was rapid chess, yes, but a serious event and the machine had played quite well in parts. Still no moon landing, but low Earth orbit had been achieved.

Both games with Genius reflected the unique nature of computer chess, especially the second game. Chess players have the most trouble visualizing the moves of knights because their move is unlike anything else in the game, an L-shaped hop instead of a predictable straight line like the other pieces. Computers, of course, don't visualize anything at all, and so manage every piece with equal skill. I believe it was Bent Larsen, the first GM victim of a computer in tournament play, who stated that computers dropped a few hundred rating points if you eliminated their knights. This is an exaggeration, but it certainly seemed that way sometimes. There is a similar effect with the queen, by far the most powerful piece. On an open board, that is, one mostly uncluttered by pawns, the queen can reach nearly every square in just a move or two. This raises the level of complexity dramatically, something computers manage far better than humans. Facing a computer with a queen and knight in an open position near your king is a horror fit for a Stephen King novel.

For all of chess history, even the greatest players had been sheltered from the sort of incredibly complex tactical play that computers handled almost trivially by 1993. You knew that your human opponents had roughly the same limitations as you did when it came to dealing with whatever arose on the board. In my case, I always felt that I had the advantage in calculation over anyone except the Indian star Viswanathan Anand, who was justly famous for his speedy tactical play. Generally, I always knew that if I couldn't be completely sure of what the consequences of my move were going to be, my opponent couldn't be sure either. That perceived equilibrium went out the window when you were facing a strong computer. It played chess well, but also differently.

The psychological asymmetry and physical factors I've already mentioned were an issue, but the new sensation of always wondering if your opponent might be seeing something you could scarcely imagine was very disturbing. It created a terrible tension in complex positions, a sense of dread that at any moment a shot could ring out in the dark. In response, you double- and triple-checked your calculations instead of trusting your instincts the way you would against a human opponent. All of this extra calculation cost you time on the clock and also made the games more physically taxing.

After a lifetime at the chessboard, you have no choice but to become a creature of habit, and those habits were all disrupted when playing against a machine. I didn't like it, but I also wanted to prove I could overcome these handicaps, and to prove that I was still the best chess player in the world, human or machine.

PC PROGRAMS were making impressive progress, but Deep Thought was not off my radar. I had had another close encounter with the IBM group in Copenhagen in February 1993, when the machine took on a Danish team that included Bent Larsen. IBM Denmark was eager to put their new employee to work. At this point the machine was Deep Thought II, but the IBM PR team had decided to call it Nordic Deep Blue in Copenhagen, apparently to distinguish it from the further-upgraded version they were building to challenge me at some point in the future. But I believe I'll avoid more confusion than I'll cause by simply referring to it as Deep Blue from now on.

Whatever they called it, the machine they brought to Denmark did not impress me. We used it to analyze one of my games for the audience, curious to see what suggestions it might have. Its evaluations of the game were poor, consistently underestimating my attacking chances, and it would only slowly realize that its proposed improvements didn't work. Still, it did capably well against Larsen and the other Danes for a performance rating of nearly 2600, and I was made aware that great improvements were in the IBM pipeline. The founding teammates Feng-hsiung Hsu and Murray Campbell had added Joe Hoane as a programmer, not to mention a sizable team and resources back at IBM, where the Deep Blue team would soon be moved to the company's premier research facility in Yorktown Heights, New York.

IBM had a new CEO, Lou Gerstner, who had come in during a very low point for the eighty-year-old company. IBM's stock had plummeted as the company struggled to keep up with a plethora of nimble new competitors. Among other things, Gerstner put a stop to a plan to dismember IBM into separate companies, which might have put an end to the chess project altogether.

In May 1995, I was able to get revenge on Chess Genius in a rapid match on German TV back in Cologne. I suppose it's a little silly to talk about revenge against a piece of software that may as well be counting grains of sand, but it felt good nonetheless. The first game should have ended in a draw, but Genius caught a case of that old chess machine disease, excessive greed, and came under a decisive attack against its king after it grabbed a distant pawn. The second game I drew with black without somersaults. In the interview afterward I confessed that I had been practicing at home against a version of the program in order to be as prepared as possible.

At the end of the year, I played another mini-match, this time against Fritz 4 in London. The constantly increasing version numbers were beginning to be a little intimidating, honestly. Perhaps I should have insisted on being called "Kasparov 6.0" after I won my sixth world championship match. It's not too far-fetched, considering that a PC program called "Kasparov's Gambit" came out in 1993, published by the American software giant Electronic Arts. It had a strong engine, colorful graphics, and occasionally a little video of me would pop up with basic advice on how the game was going. "Watch your pawn!" or "You're not on the right track now." It felt very cutting-edge at the time, but I'd probably laugh if I could find a working version now.

One of the interesting things about following the evolution of the PC programs from one version to the next was that I could always detect the programs' DNA, as it were. There would be new code added, new search algorithms and optimizations for a new generation of processors, but for the lack of a better word the damn things had style. I joke about the programmers treating their machines like children, or at least pets, but there is no doubt their creations take after them in some ways, and that these characteristics are passed down from one version to the next like green eyes or red hair. The traits weaken over time as well, as you would expect in any hereditary system.

For example, Fritz was infamously materialistic, always keen to grab a pawn and hold on to it for dear life no matter how ugly its position got. This isn't to slight its programmer, Morsch, at all, but the soft-spoken Dutchman would be the first to admit his program was never one of the most aggressive on the market. Then you had the program Junior, a winner of many championships and created by the Israeli duo Shay Bushinsky and Amir Ban. It was revolutionarily aggressive, readily giving up material for open lines and attacking chances in a way that could only be described as completely uncomputerlike at the time. Is it going too far to wonder if the stolid Dutch-German program and the fiery Israeli engine had absorbed some of their stereotypical national characteristics? Well, probably so, but a program taking on the personality of its programmer is quite natural, especially if the programmer is a strong enough chess player to appreciate the stylistic qualities of his creation.

The genetic fingerprints of the different engines were also a practical matter for me and the other Grandmasters who were battling them at the board for a decade or so of competition. You couldn't expect to practice against the exact engine you were going to face in a tournament or match, but even having an older version, or at least as many of its past games as possible, made a big difference in preparing for them. As the machines accumulated a track record over years of human-machine and machine-machine play, we could prepare for them much like we would for our Grandmaster peers. There was always the problem that they could adopt completely new openings or even a new "personality" between events, or even between games, but rarely did they change completely, although they did keep getting stronger.

The two London rapid games against Fritz 4 were only memorable because of another unique aspect of playing against a computer. On my seventh move with the black pieces I played my bishop two squares, from c8 to a6, to use standard algebraic board notation. But the human operator of Fritz wasn't paying close attention and thought I had placed it one square short, on b7, and entered that move instead.

Incredibly, the game progressed for four more moves before the operator noticed his mistake. Even more incredibly, the game was actually playable when the bishop was then placed on the correct square in the computer, although of course it would have played quite differently. I won that game and then drew the second to take the match, though it was hardly satisfying after the bizarre blunder. At least Fritz wasn't capable of being annoyed at its human handler for getting it into trouble.

IN EARLY 1995, at long last, there were inquiries by David Levy and Monty Newborn about the possibility of a match against Deep Blue, probably the next year, and I told my agent Andrew Page to keep an eye on it. When I had met their team in Denmark two years earlier, I had joked that they had to hurry up and get it ready because I wanted to face it while I was still young and strong, since I was about to turn thirty at the time. And as confident as I always was in my immortality, I wouldn't be the world champion forever. IBM wanted the match and so did I; the question was whether Deep Blue would be ready.

Hsu's compulsive perfectionism with his chess chips kept pushing back deadlines, although as a fellow obsessive I can only sympathize with him. If any one small category of people did more to build the American century than any other, it was gifted engineers who had big dreams and followed them, come hell or high water. But the parts of the machine that were working were always having problems. Reading Hsu's and the many other accounts of Deep Blue's development and play during 1994–95 quickly starts to sound like the diary of someone from the Geek Squad repair company. Bugs, crashes, disconnected phone lines, interrupted Internet connections, opening book errors, more bugs, loose circuits—everything but a virus. Meanwhile, IBM still wanted the machine on the road playing in tournaments and exhibitions for its PR value.

One of these events was the 1995 World Computer Chess Championship in Hong Kong. Deep Blue Prototype—as it was called this time, though apparently still the same basic machine as Deep Thought II since the new hardware still wasn't ready—was the big favorite. It hadn't lost to another machine in a tournament in years and, according to Hsu, it beat the top commercial programs by a three-to-one margin in their tests. (It was a big advantage that they could test against many of their competitors simply by buying a copy of the engine, while no one else could test against them.)

But as the saying goes, upsets can happen and that's why the games are played. Deep Blue drew its fourth game against a PC program named WChess and would play Fritz 3 in the fifth and final round. Deep Blue had a half-point lead and, again according to Hsu, "It won about nine out of ten games against Fritz in our pre-tournament tests back at IBM," and would have the advantage of the white pieces. Fritz played a sharp line of the Sicilian and got a fine position when apparently Deep Blue got tricked by a transposition of moves and was out of its own opening book, thinking for itself.

If Deep Blue were really so much stronger than Fritz, that shouldn't be considered much of a problem. To be fair, however, the opening was indeed a difficult one that even a modern computer could have trouble navigating out of book. Deep Blue was like the junior players I criticize in my coaching sessions for blindly following opening theory and then having no understanding of the position once their memorized lines end. Still, looking at the game, it wasn't as bad as all that. A player with an estimated 200-point rating advantage shouldn't have too much trouble holding such a position.

But out comes the Geek Squad again! Deep Blue's connection between Hong Kong and New York was lost and the entire machine had to be restarted and reconnected. According to Hsu, this "cold" restart set its thinking back and it chose a different move than the one it had been considering before the disconnection occurred.

Before moving on to the exciting conclusion of this little machine versus machine drama, I want to draw attention to what just happened because it is relevant to my own encounters with Deep Blue. In nearly every Deep Blue game description I can find from this period there are resets, crashes, reboots, and disconnects. It had to forfeit a game in a Harvard competition due to a power failure. It resigned a game against women's world champion Xie Jun in Beijing due to crashes. But this is the nature of experimental technology being assembled in haste, and rules are usually in place to handle such eventualities.

The crashes themselves don't overly concern me, but two other things about this incident do. The first is that operator intervention is required to get the machine back into the game. This wasn't just reconnecting the phone line or waiting for the Internet connection to come back online. Input was required—"We had to restart Deep Thought II," writes Hsu—and, I assume, the entire game had to be entered into the machine before telling it to start playing again. As a logical consequence, Deep Thought made a different move than the one it had favored before the crash. Hsu again: "According to Joe [Hoane], who was watching the game from our lab in Hawthorne, Deep Thought II did switch to an alternative move. But the new move never showed up on our screen in Hong Kong before the line drop, and we did not know about it until after the game."

For the sake of argument, let the assumption stand that the move Deep Blue was considering before the crash was superior to the one it played. (Looking at the game now I can say that, yes, its post-disconnect thirteenth move was indeed unlucky.) Unfortunate, of course, but what if the new move had turned out to be stronger instead of weaker? The vagaries of computer chess thinking being what they are, it's perfectly plausible that the machine might have taken more time after a reboot and found an improvement, or simply made a different move quickly that turned out to be better; who can say? The implications of this are alarming, even if you want to be charitable.

The game continued with a big advantage for Fritz. In an unfortunate attempt to defend Deep Blue's honor, Hsu's book provides commentary on the rest of the game that is complete nonsense. I may not know much about the "0.8-micron CMOS process" or the other things that make Deep Blue tick, but I still know a few things about chess. He writes about "muddling along" and "not busted yet" as if the game was competitive. In reality, although it apparently wasn't aware of it at the time, Deep Blue was completely lost after making two more terrible moves soon after the disconnection. The first blunder, its very next move, actually went unpunished when Fritz missed a crushing response. Two moves later, already losing badly, Deep Blue committed suicide by overlooking the power of black's kingside attack. It was over. Both the 3000-rated engine on my PC and the 2800-rated engine in my skull can see at a glance that white is dead meat after Fritz's sixteenth move. With nothing to lose, Deep Blue played on down a huge amount of material before finally resigning on move thirty-nine. In a huge upset, the little German David had downed the IBM Goliath and went on to win the world championship.

I was happy for Frederic and my friends at ChessBase, but this result had the potential to be a little awkward for any match against me down the line, since Deep Blue wasn't the world computer chess champion and the next championship was likely years away. In the end, this turned out to be nothing. There wasn't really any doubt that Deep Blue was still the strongest chess machine around, especially since the version I would face in Philadelphia nine months later would finally be the upgraded one that was far stronger than the machine that lost to Fritz in Hong Kong.

Meanwhile, there was the little matter of making sure that *I* was still a world champion. My 1995 title defense was a twenty-game match against India's Viswanathan Anand. We played in New York City, on the 107th floor of the South Tower of the World Trade Center. The ceremonial first move of game one was made by Mayor Rudy Giuliani, and the date was 9/11.

I'LL SHARE some details on that human versus human event later, and how a machine helped me retain my title, but one opponent at a time. February 10, 1996, would become yet another addition to my dubious collection of This Day in History dates. Previous to sitting down against Deep Blue in Philadelphia for that first game of our six-game match, I had been the first world champion to lose a blitz game to a computer and the first world champion to lose a rapid match to a computer. The trend was clear. By the time I sat down across from Hsu for game one, I understood that eventually, if I kept my title long enough, I would also become the first world champion to lose a classical game and match against a computer. But I was not ready for it to be today.

The match was sponsored and hosted by the Association for Computing Machinery (ACM), which had a long involvement in computer chess. They were celebrating the fiftieth anniversary of the first digital computer, the ENIAC, at their annual Computing Week event in Philadelphia. Monty Newborn, a chess programmer himself, had used his post at ACM to become an impressive promoter of human versus machine chess. As an intermediary between the parties, he helped work out the rules of the Philadelphia match, heralded as the ACM Chess Challenge. The International Computer Chess Association (ICCA) was the sanctioning body for the match and ICCA vice president David Levy assisted with the negotiations and organization. The prize fund was $500,000, with $400,000 going to the winner. The 4–1 split was a compromise after my counterproposal of a winner-take-all match instead of the 3–2 in the original proposal. I was very confident, and, after over six years of waiting since I defeated Deep Thought in 1989, it was fair to think that they needed me more than I needed them.

A few other factors made that not entirely true, however. Intel was dropping its support of my fledgling Professional Chess Association and its Grand Prix tournaments and I was hoping to establish a similar partnership with IBM. My dramatic and ill-advised breakaway from the World Chess Federation (FIDE) in 1993 had made me even more of a lightning rod in the chess world, but by bringing in new sponsors with the PCA we were organizing great events and putting money into the pockets of many players. But Intel Europe informed us that they weren't renewing the agreement. One of the reasons I played the Philadelphia match and the New York City rematch for less than the million dollars I thought they were worth was in the hope of establishing a long-term sponsorship arrangement for the PCA with IBM.

Predictions around the long-awaited match were very much in my favor. David Levy boldly predicted a 6–0 sweep for me. IBM's team leader C. J. Tan and I both predicted a 4–2 victory—him for Deep Blue and me for myself. I was confident, but worried about the lack of information available about this new version's capabilities—not the technical specifications, which were useless to me, but what mattered to a Grandmaster's preparation: games. The version I was facing had never played publicly before, so I really had no idea what it was capable of.

Certainly the numbers were impressive. The previous model, the last officially to be called Deep Thought, searched between three and five million positions per second. This new one, with its 216 new chess chips connected to an IBM RS/6000 SP supercomputer, reached one hundred million. I knew that twenty times faster didn't mean it was twenty times better, but it was still a black box to me and that is never a pleasant feeling. According to the experts, the "speed to depth to strength" formula that had held steady in machine chess for decades might put this new version at over 2700. A better opening book and more chess knowledge might add another 50 or 100 points, approaching my 2800+ level. But this was all theoretical. And who knew what other tricks it might have up its sleeve?

Along with all the hardware and software improvements, Deep Blue had acquired an important new teammate, American Grandmaster Joel Benjamin. The debacle with the opening book in Hong Kong had convinced the IBM team that they needed professional help, so they hired a GM to prepare the opening book and to perform as Deep Blue's second during the match in case any book adjustments were needed. Benjamin would also play a role as the machine's sparring partner and in tuning its evaluation function. Even the fastest chess machine in the world needed a little human chess knowledge.

I was also taking the match seriously. I flew into Philadelphia from Rio de Janeiro, where I had just beaten a strong Brazilian team in a simultaneous exhibition. I arrived with my own second, my trainer Yuri Dokhoian. My mother, Klara, also attended, making sure all the conditions were correct in the playing hall, and was always seated in the front row. Frederic Friedel was there to serve as my unofficial computer chess advisor. Ken Thompson, the creator of Belle and who was still very involved in computer chess, agreed to be a sort of neutral overseer of the computer. Compared to the circus that the rematch would become in New York City a year later, this first match seemed almost quaint. A considerable media presence built up as the match drew more attention, with journalists from most major print publications and even regular coverage on CNN. But it still felt relatively casual and open there in the giant convention hall. With ACM and the ICCA running the show, IBM had a relatively discreet presence, usually through the team leader, C. J. Tan. It all felt very much like any other top-level chess match, right up until the moment I sat down at the board with Deep Blue for the first time.

I'VE HAD twenty years to come up with a good way to describe what it's like for a world champion chess player to play against a world-champion-level chess machine. I'm still not sure I've succeeded. Directly competing against a computer at the highest level of a human discipline is a unique experience. It's not a video game against a computer AI or a metaphorical competition in the job market, the "race against" or "race with" the machines explained so capably by MIT's Erik Brynjolfsson and Andrew McAfee in their books.

John Henry competed with a steel-driving steam engine before a crowd of witnesses, his muscle and bone versus the implacable iron beast. Jesse Owens's races against cars and motorcycles also boasted that same tragicomic asymmetry; it was exploitation and entertainment, not serious competition. If a person wins a footrace against a car, it's funny. If he loses, what else could anyone expect?

The other difference was apparent from the popular news coverage, which echoed centuries of romantic notions about chess and intelligence, and misconceptions about artificial intelligence and Deep Blue. "The Brain's Last Stand," "Kasparov Defends Humanity," "The machines are entering the last human refuge, intelligence." Even the jokes about the match on shows like Jay Leno and David Letterman had a nervous, slightly apocalyptic feel to them. "Kasparov looks pretty nervous. You may think this is no big deal, but wait until that thing comes for *your* job!" "He's playing chess against a supercomputer and I still can't program my VCR!" "In a related story, earlier today the New York Mets were defeated by a microwave oven."

For the most part, the more flattering narratives were indulged by the organizers—and by the participants, I admit. Who was I to say that chess was not the "pinnacle of human intellectual activity"? Or that I wasn't a "living Mount Everest" or potentially a "chess champ guilty of letting down the whole human race"? And IBM had no incentive to disagree with any assumptions about its machine's "creativity" or "potential to revolutionize entire industries." ACM's Monty Newborn was in his element. He is a natural raconteur who never lets his background in computer science and chess get in the way of a P. T. Barnum sensibility. I didn't have much time for such things at the time, but now even I am almost inspired by listening to Newborn talk in interviews at the match about "what it means to be human" and likening a potential Deep Blue win to the moon landing.

Finally, all the hype and mythologizing could be put aside and the first game could begin. At least, it could after another bug was squashed by the operator. To my amazement, Deep Blue wasn't running yet when the arbiter started its clock, and it took Hsu, the operator that day, a few minutes to get it going. It may sound petty to speak of such things as distractions, but of course they are. It's difficult enough to summon your usual focus under such strange conditions, especially when you know your opponent has no such concerns. The scrum of photographers around the table doesn't annoy a computer. There is no looking into your opponent's eyes to read his mood, or seeing if his hand hesitates a little above the clock, indicating a lack of confidence in his choice. As a believer in chess as a form of psychological, not just intellectual, warfare, playing against something with no psyche was troubling from the start.

After a few moments Deep Blue was running and Hsu made its move, 1.e4. That's moving the pawn in front of the king forward two squares, and I answered with 1..c5, my favorite Sicilian Defense, a sharp counterattacking opening. Don't worry, I'm not planning on presenting the whole game! It's one of the most famous games in chess history and there is plenty of analysis of it available if you are interested. Unfortunately, it's not a very good game, as I rediscovered when I went over these matches again. To help me maintain objectivity, a couple of strong players in Moscow also went over them with the best chess engines currently available. I did not play very well in Philadelphia, even if I did play well enough.

Deep Blue declined my challenge of an open game, somewhat surprisingly since computers thrive in the complex tactical positions that the Open Sicilian is known for. The IBM team was worried about running into an opening surprise I might have cooked up and clearly didn't feel like it was wise to match Joel Benjamin's preparation against mine in a risky variation. Instead it played the same second move it had used against me with white in our 1989 match, although of course they wouldn't expect me to repeat that game despite the result. Trying to repeat your past victories without having improvements prepared in your own play is a very good way of walking into a preparation land mine. They instead made a solid choice and the machine was well prepared, reaching the ninth move still in its opening book.

I was also prepared, and varied from a previous game of mine on the tenth move with an improvement. I wasn't going to go into a defensive crouch. I wanted to see what this thing could do. This wasn't a blitz or rapid game; we had hours on our clocks, not just a few minutes. This gave me enough time to think, so I wasn't afraid of entering sharp complications. Deep Blue played well in the early stages, gaining a slight advantage typical of the white pieces. After I made an inaccuracy it played several more strong moves to create real threats for the first time. I glanced up at Hsu, a habit rendered pointless in this match. My position was deteriorating. This thing was strong. This was different.

READING THROUGH some of the dozens of books and hundreds of articles written about this match, and this game in particular, you would think the authors had all attended different events and analyzed a different game. Disagreements in analysis are normal, of course, and healthy. When someday chess has been completely solved by some technology we cannot even imagine now, we will be able to speak of objective truth on the board. Until then, we will have regular disagreements about the quality of some moves. Different Grandmasters and different machines will prefer different ideas that may be equally strong, which is what makes chess interesting.

This isn't to say that some moves aren't simply blunders or inaccuracies, or that often there isn't a clearly best move on the board. In many positions the right move is obvious and will be made by any reasonably strong player. Perhaps 10 to 15 percent of positions require a master's experience or calculation skills to find a sophisticated plan or complicated tactic. Then there is that last 1 or 2 percent of moves, the very difficult ones that even strong Grandmasters might miss. Under such conditions, with the stress of competition and the pressure of the clock, it's remarkable that humans play chess as well as we do. In fact, I discovered that it is often the case that we perform better under pressure, not worse.

During my work on the *My Great Predecessors* book series, I gained not just a deeper respect for the achievements of the past world champions I was studying, but a greater admiration for chess players in general. Few activities are as taxing to the human faculties as a game of professional chess. Rapid calculation is essential, adrenaline is surging, and the outcome hangs on every move. This goes on for hour after hour, day after day, often with the whole world watching. It is the ideal scenario for a mental and physical meltdown.

When I began to analyze the games of my world champion forebears, I was therefore prepared to be a little forgiving. Not in my analysis, where I had to be as merciless as my teacher Botvinnik had instructed me, but in the tone I adopted toward their mistakes. Here I was in the twenty-first century, with databases of millions of games and gigahertz of chess engine processing power at my fingertips. "With these advantages and the benefit of hindsight I shouldn't judge my predecessors too harshly," I told myself.

An important part of the project was to collect all the relevant analysis that had been done on these games before, especially the published analyses of the players themselves and their contemporaries. My colleague Dmitry Plisetsky did a phenomenal job of tracking down sources in a half-dozen languages. One would assume that the analyst, working in the calm of his study and with unlimited time to move the pieces and make notes, would have a much easier job than the players themselves. Hindsight is 20/20, is it not? But one of my first discoveries was that when it comes to chess analysis in the precomputer age, hindsight was badly in need of bifocals.

Paradoxically, when other top players wrote about games in magazines and newspaper columns they often made more mistakes in their commentary than the players had made at the board. Even when the players themselves published analyses of their own games they were often less accurate than when they were playing the game. Strong moves were called errors, weak moves were praised. It was not only a few cases of journalists who were lousy players failing to comprehend the genius of the champions, or everyone missing a spectacular move that I could easily find with the help of an engine, although that did happen regularly. The biggest problem was that even the players would fall into the trap of seeing each game of chess as a story, a coherent narrative with a beginning and a middle and a finish, with a few twists and turns along the way. And, of course, a moral at the end of the story.

I took two lessons away from this discovery. The first is that we often do our best thinking under pressure. Our senses are heightened and our intuition is activated in a way that is unique to stress and competition. I would still rather have fifteen minutes on my clock than fifteen seconds to make a critical move, but the fact remains that our minds can perform remarkable feats under duress. We often do not realize how powerful our intuitive abilities are until we have no choice but to rely on them.

The second lesson was that everyone loves a good story, even if it flies in the face of objective analysis. We love it when the most annoying character in the movie finally gets what he deserves. We root for underdogs, cringe at a hero's downfall, and sympathize with the unlucky victim of the Fates. All these tropes are in play in a chess game, just as they are in an election or the rise and fall of a business, and they feed the powerful cognitive fallacy of seeking a narrative where often none exists.

Computer analysis exploded this lazy tradition of analyzing chess games as if they were fairy tales. Engines don't care about story. They expose the reality that the only story in a chess game is each individual move, weak or strong. This isn't nearly as fun or interesting as the narrative method, but it's the truth, and not just in chess. The human need to understand things as a story instead of as a series of discrete events can lead to many flawed conclusions. We are easily drawn away from the data by a nice anecdote that fits our preconceived notions or that fulfills one of the popular tropes. This is how urban legends propagate so efficiently; the best ones tell us something we really want to believe is true. I'm certainly not immune to this tendency myself, and it's impossible to overcome all our intellectual biases. But becoming aware of them is a good first step, and one of the many benefits of human-machine collaboration is helping us overcome lazy cognitive habits.

With all that in mind, let's return to the board, where I was getting into real trouble in my first game against Deep Blue. The machine had played several surprising moves and I was accumulating weaknesses in my position. Looking over the analysis of others and listening to the commentary that was being provided live by several Grandmasters (and Fritz 4!), the tendency toward narrative has overwhelmed objectivity once again. The consensus seems to be that I made the fatal mistake of counterattacking a computer in an open position where its unmatched tactical abilities would be overwhelming instead of trying to consolidate and sit tight. Perhaps this is true, but it was not my intention to play to the computer's strength. I simply did not see a better choice.

After my 1989 victory over Deep Thought, I was interviewed by the *New York Times* for a lengthy magazine article. We were looking over the news coverage the match had received and one quote from Deep Thought team member Murray Campbell caught my eye. "Deep Thought didn't get a chance to show what it can do," he said. "That's exactly the point!" I shouted to the interviewer. "I didn't let it! The highest art of the chess player lies in not allowing your opponent to show you what he can do."

Seven years later, Deep Blue was proving too strong to so easily dictate terms to, especially since it had the white pieces. And while my choice to attack its king can be criticized as ill advised against a machine, it was not a bad move, and certainly not the losing move. That would come two moves later when, ironically, I held up my attack to preserve a pawn. Had I continued to play as aggressively as all the commentators criticized me for doing, I might have saved the game. But that would go against the popular narrative, so the losing move is often overlooked.

What I overlooked, on the other hand, has been correctly diagnosed. Deep Blue grabbed a pawn far from the action in what appears to be a terrible loss of time with its king under attack. But, in the time-honored tradition of human-machine chess, it had calculated deeply enough to get away with it. Despite what I've said about the dangers of narrative, I cannot resist sharing this passage on the game from Charles Krauthammer's story on the match for *TIME* magazine. This sort of storytelling I completely endorse.

> Late in the game, Blue's king was under savage attack by Kasparov. Any human player under such assault by a world champion would be staring at his own king trying to figure out how to get away. Instead, Blue ignored the threat and quite nonchalantly went hunting for lowly pawns at the other end of the board. In fact, at the point of maximum peril, Blue expended two moves—many have died giving Kasparov even one—to snap one pawn. It was as if, at Gettysburg, General Meade had sent his soldiers out for a bit of apple picking moments before Pickett's charge because he had calculated that they could get back to their positions with a half-second to spare.
>
> In humans, that is called sangfroid. And if you don't have any sang, you can be very froid. But then again if Meade had known absolutely—by calculating the precise trajectories of all the bullets and all the bayonets and all the cannons in Pickett's division—the time of arrival of the enemy, he could indeed, without fear, have ordered his men to pick apples.

> Which is exactly what Deep Blue did. It had calculated every possible combination of Kasparov's available moves and determined with absolute certainty that it could return from its pawn-picking expedition and destroy Kasparov exactly one move before Kasparov could destroy it. Which it did.
>
> It takes more than nerves of steel to do that. It takes a silicon brain. No human can achieve absolute certainty because no human can be sure to have seen everything. Deep Blue can.

I held out my hand to resign on move thirty-seven and a computer had defeated the world chess champion in a classical game for the first time in history. I was a bit in shock, as were the spectators and commentators. Even Hsu, who would have been aware of Deep Blue's winning evaluation from his screen, looked a bit confused, almost apologetic on the moment of his great triumph. I honestly feel a little bad about that now, despite the bad blood that would arise out of the rematch a year later. I'm sure he wanted to jump up and down with his teammates, not answer my questions.

Still in a mild daze at how well the machine had played, I asked a reflexive question immediately after resigning, the way two Grandmasters might begin what we call the "postmortem" of a completed game. "Where did I go wrong?" I asked. But Hsu wasn't much of a chess player and, probably a bit dazed himself, he couldn't recall enough of Deep Blue's analysis on the screen to answer, so it was a slightly awkward moment for both of us.

A month after the match, I wrote in *TIME* that I felt I could sense "a new kind of intelligence across the table" that day, and in a way, it was true. I wasn't suggesting any metaphysical interpretation, but could sheer speed really produce such impressive chess? Several of its moves were almost as if it was saying, "I bet you didn't think a computer could make a move like this!" For example, at one point in the middlegame it sacrificed a pawn for activity, a very humanlike idea not at all in keeping with the usual machine materialism.

It was the best I had ever seen a machine play, against me or anyone else and, at least at the moment of my loss, I even considered the possibility that it might be too strong to beat. Later that day, I wondered aloud to Frederic, "What if this thing is invincible?" I had known that day would arrive eventually; was it here already?

I didn't have to wait very long for the answer. In game two the next day, I played a slow, maneuvering opening with white. The idea was to not provide Deep Blue with any clear targets, knowing it couldn't formulate strategic plans the way a human could. At least, I hoped it couldn't. As usual, there were a few technical problems, although I was only aware of one of them at the time. Deep Blue played a poor move very early, on move six. According to Frederic, I was visibly delighted by what I could only assume was a major flaw in the machine's opening book. Not only wasn't it invincible, I was going to have an easy day of it. You can imagine my disappointment when the arbiter ran over to say that Hsu had accidentally made the wrong move on the board, capturing the wrong pawn, as had happened in my London match with Fritz. The rules allowed them to correct it and the game proceeded along normal lines. It all turned out fine, but it illustrated the danger of having a weak player making the moves and how distractions like this only affected the human player.

Hsu's book blames Murray Campbell for failing to properly upload the updated opening book file he and Benjamin had worked on after game one to the machine back in Yorktown Heights. This left it to rely on something he calls an "extended book," which had vague guidelines based on database statistics from Grandmaster games. Regardless, I was oblivious to this and Deep Blue played the opening just fine, following high-level Grandmaster theory until I introduced a new idea on move fourteen. Several books also mention an "evaluation bug" in Deep Blue that affected its play in this game, but honestly I tire of trying to figure out which bugs are "bugs" and which "bugs" are bugs, and which are just lousy evaluations.

My strategy worked out quite well and Deep Blue was saddled with the sort of long-term structural weakness it had no idea how to defend. I realized that just avoiding wild tactical positions wasn't enough. I should aim for positions in which general principles would outweigh short-term calculations. Deep Blue did have evaluation functions, but it was not very sophisticated and something I could exploit once I became aware of its hard-coded preferences. For example, if I noticed it had been set to keep queens on the board—generally a good idea for a machine against a human—I could play moves that offered it the choice of exchanging queens or making an inferior move.

This sort of human adaptation was one of the reasons some computer scientists thought that chess machines wouldn't defeat Grandmasters for far longer than turned out to be the case. Once a human figures out the rules and knowledge that govern a machine's play, they thought, they would figure out how to exploit them. But it turned out that with super-fast brute force, little exploitable knowledge was needed and most weaknesses were amply covered up by sheer depth of search.

Deep Blue hadn't achieved perfection just yet, however. In game two, I offered a pawn sacrifice that it couldn't resist and in compensation the light squares around its king were fatally weakened. It was close to reaching a draw, but the best lines were always a little too deep for its search and it didn't know the general principles of how to defend such positions. After hours of careful maneuvering I won one pawn and then another and Murray Campbell resigned for Deep Blue on move seventy-three. I had leveled the score and, more importantly, I knew it was only mortal.

Now that I knew that the "new kind of intelligence across the table" was only a much faster version of the computer programs I understood well, I relaxed a bit. It was very strong, yes, but it wasn't stronger than I was and it had clear deficiencies. As with a human opponent, if I could target its weaknesses and avoid its strengths, I would win the match.

Game three repeated the opening from the first game until Deep Blue deviated with a move inserted into its book that day by Benjamin. We continued along the line he planned until move eighteen, when Deep Blue noticed that the line Benjamin had intended, but lucky for him, not inserted into the book, actually lost a piece. This left me with a small advantage and a clear target to focus on, so I thought my chances were good for a second consecutive win. But Deep Blue started to do what machines are known for, impossibly tenacious defending, harder to kill than a real bug, a cockroach. If there is only one move to save the position, they always find it. Much to my frustration, Deep Blue found a long sequence of resourceful moves to escape danger and draw the game.

Precision under fire is another aspect of human versus machine asymmetry. What we call a "sharp" position in chess is when there is high complexity and grave consequences for any error. Both players are balancing on a tightrope and the first slip can be fatal. For a computer, this actually makes it easier to find the right path because all the other moves return a very low score. Humans can never enjoy such confidence. What's more, only the human player is aware that there is a tightrope. I can sense danger in a position, feel the tree of variations growing exponentially. That's just another day at the beach for a machine, especially one that has, as Deep Blue did, special search extensions that added extra depth in consequential variations.

The match was level after three of the six games, but I had white in two of the final three and was feeling more comfortable. The media attention for the match had grown tremendously after Deep Blue's win in game one, but of course the machine didn't have to give interviews. I disappointed my computer expert by ignoring his advice and opening the position in game four. I didn't shy away from playing sharply with white. I pondered a piece sacrifice against Deep Blue's kingside on move thirteen for a while before deciding it was simply too risky. It's notable, however, that I would have played it against any other chess-playing entity on the planet, man or machine. I knew that if I made the slightest miscalculation in such a position I was dead and would be behind in the match with only two games to go. It was an important moment, in retrospect. I wasn't just playing chess, I was making specific adjustments to playing against a machine whose capabilities in certain areas far exceeded mine or anyone else's.

There was yet another technical snafu during game four, and it came exactly when I was preparing a dangerous attack. I had spent a long time on my previous move, planning to sacrifice a knight for two pawns and an attack. Before Deep Blue replied, it crashed and had to be restarted. I was furious, ripped out of my state of deep concentration at a key moment in the game. It took twenty minutes to get it working again and when it came back it played a strong move that avoided my sacrifice. It was enough to make me wonder if something more than bugs was going on. (Subsequent analysis shows the sacrifice would likely have led to a roughly equal position.)

The position was now balanced, but sharp, and I was approaching time trouble. If I reached move forty, more time would be added to the clocks; the question was whether or not I would make it. After making several precise moves I got to the safe shore of the time control at move forty with a defensible position. I found a nice way to force a drawn position and the game was soon over. The score was still level with two games to go and I was exhausted. Attendance at the match had continued to climb and the media attention was approaching a frenzy. There were interviews and TV appearances for both teams, and IBM definitely noticed that their little chess project was getting more attention than just about anything else they had done in years.

Despite the rest day between games four and five, I had trouble mustering my energy. I avoided my usual Sicilian for the Russian Defense, a.k.a. the Petroff. This wasn't a display of patriotism; the Petroff is very solid, some would say boring. It often leads to many piece exchanges and symmetrical pawn structures that reduce the dynamism in the position, something I thought sounded ideal when tired and facing a supercomputer, even though it wasn't the sort of position I usually played. Deep Blue transposed instead into a Four Knights Opening, which was no more or less as dull as the Petroff.

After many exchanges took place I had the tiniest of advantages. Thinking of saving my energy for the final game with white the next day, I offered an early draw on move twenty-three. For those new to the chess world, the idea of offering a draw must sound very strange. Imagine two boxers simply agreeing to stop fighting in the second round, or a soccer match ending after fifteen minutes because the coaches decide a tie is a good result. Typically, until rules were put in place to discourage it, in chess either player can offer her opponent a draw after any move. The other player can then think about it and accept, or ignore the offer and make a move, and the game continues.

Draws have always been a part of chess, at least in the modern history of the game. There are many positions that cannot be won by either side, including stalemate, in which the side to move has no legal moves and so the game is drawn. Draws are worth half a point for both players, so it's definitely better to draw than to lose and get nothing. The draw offer was created as a courtesy so strong players would not have to wear themselves out playing tedious and obviously equal positions all the way down to nothing. It was a way of saying, "I know you know how to draw this and you know that I know, so let's shake hands and retire to the smoking room." It may have disappointed some spectators to end the game early, but there weren't usually very many spectators to worry about. Additionally, back in the nineteenth century, the level of play was relatively low and almost all the games ended decisively.

The problem began when masters began to exploit the draw offer strategically, or even tactically. If a draw suited you in the tournament standings, why not see if your opponent would also like a short day's work and offer an early draw? Or if you felt that your position was deteriorating, perhaps offer a draw and see what your opponent thought about it? Soon enough, it became something close to a plague, with perfunctory games as short as a few minutes and a dozen moves, even between strong Grandmasters. The habit was contagious and today it's not unusual to see short draws even at the weak amateur level.

Eventually, organizers of top tournaments decided they no longer wanted to support such behavior and instituted rules like move minimums. Now it's fairly standard to have events where it is not permitted to offer a draw before move thirty or forty, although little can be done about draws by repetition of position. With players becoming stronger and more accurate decade by decade, the number of draws has increased at the top level, with roughly half the games at elite events finishing drawn. I don't see this as a problem as long as they are fighting games—a draw is a fair result. But there are regular pushes to introduce more rule changes to encourage more aggressive play and produce more decisive games, such as awarding three points for a win and one for a draw, as is used in many professional soccer and hockey leagues.

In match play, short draws can be strategically useful. I was still feeling exhausted in game five and also felt that there wasn't a great deal to play for in the position when I offered the early draw. It would have been a disappointment for the seven hundred or so spectators that day, however, so it was their good fortune that the Deep Blue team declined my offer and decided to play on. As an aside, this is another unique aspect of machine play, when to offer or accept draws. Should the decision be left up to the machine somehow? For example, if its evaluation is at zero or worse, should it automatically accept? But what if it is in a must-win situation? As with opening books, it's a case where there isn't a very good solution to what amounts to human intervention.

Deep Blue thought it was a little worse at the time of my draw offer. The team huddled and eventually followed Benjamin's recommendation that it was too early to end the game, especially since they would have black in the final game. This turned out to be my good fortune as well, as Deep Blue's next move was a serious mistake. Unable to see the long-term consequences, it stepped right into a pin that would tie up its pieces for a long time as I advanced my pawns. With no active plan, and not understanding that its only hope was to lash out, Deep Blue shuffled around for several moves. By the time the danger was close enough to reach its search horizon, it was too late to save itself. I won in forty-five moves to take the lead in the match for the first time, and was guaranteed at least a tie in the match going into the final game the next day.

I was feeling good headed into game six despite my tiredness. I had outplayed the machine in game five and felt like I was getting to know its weaknesses. This was probably an overestimation on my part after only five games, but I knew a lot more than I had a week earlier, and it would all come together in game six. We repeated the first few moves from my first two whites until Deep Blue varied. Being behind in the match, their team had the task of trying to find a way to play for a win with black, and it wasn't going to be easy. I could go entire calendar years without losing a game with the white pieces despite my aggressive style, and here I only needed a draw to win the match, and the $400,000 winner's check, so I wasn't going to take any unnecessary risks.

After my transposition of moves got Deep Blue out of its opening book it began to play weakly, and it fell into a passive position. Without its book, it didn't know, as a Grandmaster would, that certain pieces just belong on certain squares in certain openings. This is exactly the sort of generalized, analogous thinking that humans use all the time. Without it, Deep Blue had to rely on its search to keep it out of trouble, but its options were dwindling. I shoved my queenside pawns forward, driving its pieces back. It was exactly the sort of control game I had dreamed of: closed instead of open, strategic instead of tactical. I could smell blood, or whatever it had.

At move twenty-two I considered a tempting piece sacrifice against its king that looked winning. But could I be sure? Ninety percent sure, yes. Ninety-five percent, maybe. But against Deep Blue, and needing only a draw to win the match, I would have to be 100 percent sure. Analysis later showed that it was indeed a winning blow, although there is no way to guarantee I would have played it perfectly. And there was no reason for me to take any risks, since I was crushing it already. Black had no counterplay and my pawns were still on the march. The audience got quite excited when they understood what was happening. Deep Blue was being suffocated, its bishop and rook trapped on its first rank. At the end, black's pieces were so tied up that I didn't even have to break through. The machine was out of moves that didn't lose material and the Deep Blue team decided it was time to resign.

I had won the match 4–2, exactly the score I had predicted, but it had also been far tougher than I imagined it would be. I praised the Deep Blue team for their achievement. Beyond the score, it could occasionally play chess of a quality I never believed a computer could play. I adapted my strategy and won the last two games quite easily, which may not have been good for my mindset going into the rematch. I concluded my match article in *TIME*:

> In the end, that may have been my biggest advantage: I could figure out its priorities and adjust my play. It couldn't do the same to me. So although I think I did see some signs of intelligence, it's a weird kind, an inefficient, inflexible kind that makes me think I have a few years left.

In fact, I had exactly 450 days, until the end of the rematch on May 11, 1997. Looking back, I was the last world champion to win a match against a computer. Why don't those This Day in History calendars have a page for that!?

DESPITE BEGINNING with very little publicity, the first Deep Blue match became the largest Internet event in history at the time. IBM had to assign a supercomputer like the one that ran Deep Blue to handle the load on the website—and this was in 1996 when most people were on dial-up connections. It became an early example of the power of the new communications network, showing how the Internet might one day compete with television and radio. Imagine.

The Deep Blue team obviously weren't happy with the result of the match or the way the last game went in particular, but they said they were satisfied. They had beaten the world champion and made me sweat quite a bit in the first four games. Meanwhile, IBM was even happier than I was. The winner's check they gave me was nothing compared to what the match publicity had done for IBM's stock price and the company's image. Suddenly, stodgy old IBM was cool, on the cutting edge of artificial intelligence and supercomputing, battling for supremacy against the human mind. At least that's how it looked, and the stock market seemed to agree.

According to Monty Newborn's book on the match, IBM's stock rose an equivalent of $3,310 million in little more than a week, a week that the rest of the Dow Jones went down significantly. I should have demanded stock options instead of a 4–1 prize split! Deep Blue's name was everywhere in the media and the IBM team and the IBM brand went with it. It was good for me too, of course, especially in America where chess champions were hardly household names. I was getting more US media attention for beating Deep Blue in Philadelphia than I had for beating Anand in a world championship match in New York City. It turned out that even world champions are outranked by defenders of humanity.

The PR bonanza virtually guaranteed a rematch; the question was when. There was no way the Deep Blue team would want to play again until substantial improvements could be made. How long would it take for them to get a new version ready that was strong enough to be more of a threat? Because, as the negotiations went on, one thing became very clear: if there was a rematch, it wasn't going to be because the Deep Blue team wanted to improve, or because Garry Kasparov wanted another paycheck. It would because IBM wanted to win.

CHAPTER 8

DEEPER BLUE

KEN THOMPSON designed the revolutionary chess machine Belle, whose chips Deep Blue's were based on, while working at Bell Laboratories in New Jersey, the famous "Idea Factory" that did pioneering work on breakthroughs in everything from solar cells and lasers to transistors and cell phones. Thompson was also the principal inventor of the ubiquitous Unix operating system while there, which is the basis for what runs Apple Macs, Google Android, and the billions of devices and servers running Linux.

As with the early years of ARPA, the concept at Bell Labs was to describe big problems and then work on creating the technology to solve them, instead of starting with a specific product in mind. I heard similar stories when I was invited to speak at General Electric's new Innovation Center near Detroit in 2010. My hosts were eager to stimulate the sort of "blue sky" thinking that had gone out of fashion after decades of industry consolidation and acquisitions. During my seminar, someone pointed out that too often giant companies assume that even if they aren't innovating, somebody is somewhere, and when something good comes along they will simply buy it. You can see how it eventually becomes a problem when everybody thinks somebody else will innovate.

I was reminded of this particular seminar in the context of chess machines because of a slide I used with a quote from Alan Perlis, a computer science pioneer and the first recipient of the Turing Award, in 1966, awarded by ACM. In a famous list of epigrams about programming that he published in 1982, Perlis wrote, "Optimization hinders evolution." This jumped out at me because it at first sounds contradictory. How could making improvements in something prevent it from evolving? Isn't evolution itself a type of steady improvement?

But evolution isn't improvement; it's change. Usually from simple to complex, but the key to it is increasing diversity, a shift in the nature of a thing. Optimization can make computer code faster but it won't change its nature or create anything new. Perlis liked to show an "evolutionary tree" of programming languages, and how one led to another by evolving to fit needs and adapt to new hardware environments. He explained how ambitious goals are what lead to evolution because they create unexpected needs and new challenges that cannot be met only by optimizing existing tools and methods.

It's also a matter of opportunity cost. If the focus is too heavily on optimizing, nothing new is created and stagnation can result. It can be too easy to concentrate only on making something better when we might be better served by making something new, something different.

Perlis's adage can be applied broadly beyond programming, although care should be taken not to overdo it. It has itself evolved into the popular "optimization is the enemy of innovation," which ropes in another slippery term. Many things we call innovations are little more than the skillful accumulation of many little optimizations. There wasn't much new technology in the first iPhone, for example; it wasn't even the first of its kind. Nor was the iPad the first tablet, etc. But being first doesn't guarantee success, nor does being the best. Putting the right pieces together at the right time counts for a lot as well, especially in an era when marketing budgets are increasing while R&D budgets are decreasing. No invention is innately "disruptive," to use another overused term; it must be used disruptively.

Babbage, Turing, Shannon, Simon, Michie, Feynman, Thompson . . . the list could go on. With so many of the twentieth century's most important thinkers and technologists devoting so much of their time to chess, I wonder if they would have been even more prolific without it, or much less. The benefits of chess for improving concentration and creativity in kids is documented, so it's not far-fetched to imagine that the same might true for adults. Or perhaps learning chess as kids gave the brains of all these luminaries a little extra something during their formative years.

It was once believed that brain plasticity ceased before adulthood, but that consensus has been overturned in recent years. Nobel Prize winner Richard Feynman wrote extensively about how he thought his eclectic hobbies like playing Brazilian music and lock-picking actually helped him be better at physics instead of distracting him from it. Ken Thompson enjoys piloting himself around in a small plane. And even if it's too late for you to win that Nobel, it's never too late to play chess, especially since various studies now tout the benefits of games like chess and other cognitively demanding activities in delaying the onset of dementia.

Ironically, Thompson's creation of the super-fast hardware machine Belle signaled the end of the evolution in chess machines. The tremendous results acquired via speed, brute force, and optimization were too good to ignore if you wanted to make a competitive chess machine. There were still many important improvements to come in making search more efficient and adding small pieces of knowledge, but the winning concept had been found. Thanks to Internet collaboration on programming techniques, vastly better opening databases, and ever-faster chips from Intel, chess engines running on PCs were improving so quickly that the millions of dollars of custom chess chips and supercomputing power in Deep Blue would be surpassed by an off-the-shelf engine running on a business-class Windows server in just six years.

That is, at any given moment, the very best hardware-based machine you can build will be the best chess machine around. But since you need to replace all those expensive chips with smaller, faster ones to really upgrade it where it counts, hardware-based machines are frozen in time without massive continued investment. The prize of beating the world champion in a match for the first time made the investment worth it for IBM, but there wasn't much to be done with Deep Blue after that if it wasn't going to play chess, other than to send a few pieces to the Smithsonian.

I won't contradict the years of IBM press releases and interviews that justified their investment in Deep Blue by talking about it very soberly as a useful test bed for parallel processing and other IBM projects. I'm sure there is some truth to it. But I will question the *need* for their justifications. There should be nothing wrong with one of the world's great tech companies investing in a great quest, in taking part in an exciting competition that brings together pop culture and high technology. I understand that they wanted to translate the hundreds of millions of dollars in publicity our matches got for them into products and sales, but the grander message of challenge and exploration did all that and more. There cannot be a better way to capture market share than to capture people's imaginations.

DISCUSSIONS ABOUT a rematch with Deep Blue started while we were still on the stage in Philadelphia at the closing ceremony. I asked team director C. J. Tan if he thought they'd be able to substantially improve Deep Blue in the near future. He said yes, now that they understood better what would be necessary. "Good," I answered, "then I'll give you another chance!"

It was no jest, and my question to him was also serious. I knew, or thought I knew, how computers got faster over time, and how chess machines got stronger. Moore's law, speed doubling for an extra ply of search depth, each ply deeper resulting in a strength gain of around 100 points, etc. But there were clearly difficulties as well, even for a hardware machine with an experienced and talented team with the massive resources of IBM behind it. It took Deep Thought over six years to go from its roughly 2550 level to the 2700 level it had in Philadelphia. Despite its new chips, new supercomputer, and Grandmaster trainer, I had tricked it in game five and crushed it practically without resistance in the final game. Maybe the diminishing returns from search depth were setting in at the range Deep Blue was reaching? I had trouble believing they could get it up to my 2800 level without a few more years of development.

This was, I believe, a correct evaluation at the time, but there were also several problems with it. The first was how much more IBM would invest now that they had seen what a global sensation their little chess project had become. In the span of one week, the name Deep Blue had become practically synonymous with artificial intelligence, bringing IBM with it to the forefront of a hot tech sector, at least in the public eye. It was the most recognized thing to come out of IBM in years. CEO Lou Gerstner's aggressive turnaround plan included several of these high-profile projects, including using a supercomputer system like the one that controlled Deep Blue to run the networks at the 1996 Atlanta Olympic Games. Among these was a live weather forecast system quickly renamed Deep Thunder to hop on the chess machine's coattails.

If a match IBM barely showed up for until Deep Blue won the first game could do so much for the company's stock price and produce so much publicity, imagine what a rematch could do with the full power of the IBM PR machine behind it from the start. Imagine what winning a match might do. Nobody really cared that Deep Blue lost the first match, much as few remember that I won it. It was a first, organized at a convention center in Philadelphia by ACM and the ICCA, part of a scientific experiment that had been ongoing since 1948. Progress had been made with the victory in game one and Deep Blue had been the underdog. They deserved credit and they got it.

Everything would be different in the rematch and the stakes for IBM would be far higher. They were going all-in, as the poker people say, pushing tens of millions of dollars into organizing a true spectacle in New York City themselves. If Deep Blue lost again it might start to look like a waste of shareholder money no matter how much publicity it got. Instead of cutting-edge challengers, they might just look like losers. The late-night shows and cartoons would taunt IBM instead of me. Would Gerstner have the stomach to come back a third time? Maybe, but probably not so soon, and who knows what might happen over the course of a few years?

I underestimated that with so much on the line, IBM wasn't only building a chess machine to beat me at the board, but a machine to beat me, period.

The second problem in my 1996 evaluation was a loss of objectivity about my own play. As I described earlier, success can be the enemy of future success. Having beaten Deep Blue convincingly in the last two games, I made the typical and dangerous mistake of crediting my own play more than the poor play of my opponent. You might think at first this doesn't really matter since the rematch would have the same participants, but when one opponent is a machine this isn't true at all. The Deep Blue team would learn more from their losses than I learned from my wins and they would use what they learned to target my weaknesses while strengthening their own. They would address the machine's specific insufficiencies, not only double its speed.

Mikhail Botvinnik knew a few things about rematches. He became the sixth world champion in 1948 after winning a match tournament held among the world's best players after Alexander Alekhine died with the title in 1946. The USSR had produced a golden generation that would dominate chess in the 1950s and 1960s and Botvinnik was the patriarch, *primus inter pares*, first among equals. He maintained that position not by winning world championship matches, exactly, but by winning world championship rematches. He drew his first title defense against David Bronstein in 1951, holding on to the title by the rule that the challenger had to win, giving the defending champion draw odds. In 1954, he drew another match, against Vasily Smyslov. Smyslov was too much for him three years later, and Botvinnik lost the title for the first time.

Botvinnik's best move wasn't on the board, however. The rules allowed the champion an automatic rematch the following year if he lost, instead of having to go through the usual three-year qualifying cycle. The rematch clause became a useful way for Soviet political favorites like Botvinnik to improve their title odds significantly over the years. He still had to win at the board, and in 1958 he did just that, taking the title back from Smyslov after winning the first three games in a row and holding on. Two years later the cycle repeated itself. Botvinnik was overwhelmed by the dazzling chessboard magic of Mikhail Tal, the twenty-three-year-old "Magician from Riga," and lost the title for the second time by a wide margin of four points.

Few gave the fifty-year-old Botvinnik a chance in the rematch a year later, but he once again proved that underestimating the patriarch was even riskier than a Tal combination. Botvinnik dominated the rematch, winning by an even larger margin to again retake the title. He would hold it until 1963, when he lost to Tigran Petrosian, and to the rules committee that had removed the rematch clause. It was fair, but who would have bet against Botvinnik in a rematch, even against a player eighteen years his junior? Not I.

Botvinnik stayed active, establishing his eponymous school where I later became a star pupil and also investing a lot of time in writing about and developing an experimental chess program. Perhaps his greatest lesson was given in his 1958 and 1961 rematch victories over Smyslov and Tal. While his conquerors basked in a year of glory, Botvinnik spent that time doing almost nothing but analyzing the matches he lost and preparing for the rematch. He did this not only by analyzing and preparing for the play of his opponents, but with an intense regime of self-criticism. Botvinnik realized that it was not enough to find weaknesses in the play of Tal and Smyslov; he had to improve his own play and to detect and protect his own flaws. Few people are capable of such objectivity at all, and even fewer are capable of doing it as successfully as Botvinnik.

To prepare, Botvinnik focused on training matches and analysis that replicated what he believed were the games and positions he played poorly in the matches he lost. He understood that while he could not control what his opponents might work on to improve themselves, he could target his own deficiencies. Of course, it was a little different from my situation since Botvinnik was the loser in those two matches. Overconfidence could not be a problem for him, while the opposite was true for Smyslov and Tal. Still, his focus on his own play is a valuable lesson for anyone in any pursuit.

The supposedly passionless Botvinnik also found a little extra motivation by how quickly the victors were to put him out to pasture with their praise after they had defeated him. Especially Smyslov, who had written after the 1957 match about how the struggle for the world championship was finally at an end and that now Botvinnik would be free to relax a little and to play more casually without the burden of the world championship crown. Botvinnik saw opportunity in Smyslov's confidence, later writing, "Conceit does not put one in the right frame of mind for work." If only I had done a better job of remembering my teacher's words.

Had I done so, I would have realized that my play in the first match was mediocre at best, and that only Deep Blue's unique weaknesses in the final two games masked this fact. As Murray Campbell of the Deep Blue team said, I hadn't allowed their machine to show what it could do. That was partly to my credit, yes, but it also meant it gave them specific things to work on over the next year to repair those gaping holes. Unlike Feng-hsiung Hsu, Campbell had a background as a serious chess player and this made his remarks more insightful. He understood the difference between a loss and a bad loss that had to be learned from or it would be repeated. Regarding the disastrous game six, he told Newborn, "I think [Kasparov] didn't have a very complete picture of Deep Blue's strengths and weaknesses, but how can you in only five games? But I think he had enough of an idea that he stumbled on something that he was able to exploit, and it worked very well."

This is a valid perspective, even though I would give myself a little more credit than "stumbling" upon what worked. Although I had faced Deep Blue only five times, I had a great deal of insight into the general weaknesses of chess machines. They often had a poor grasp of positional factors like space—how much territory each side's forces control—a defect evident in the game six wipeout. My knowledge of machine tendencies would prove to be no substitute for specific knowledge about Deep Blue in the rematch, however, and it would even work against me when my faulty assumptions were refuted. Returning to tennis, I learned in the first match that my opponent had a lousy backhand, and I targeted that weakness. In the rematch, I expected Deep Blue to still have a lousy backhand—a poor understanding of space, in particular—but that weakness was almost entirely gone, as it showed to shocking effect in game two.

The third issue with my diagnosis of Deep Blue and the results of the first match in general was how different human and machine are when it comes to chess strength. Every Grandmaster has strengths and weaknesses. Even world champions don't play all three phases of the game—opening, middlegame, endgame—at the same level. But the range of variance in different types of positions is relatively small, and inconsistently revealed. A GM who isn't known for his endgame play might still play a beautiful endgame on a good day. Another whose openings are usually a weak spot might have prepared a devastating idea in just the line you happen to play in your game. The most gifted tactician can have a moment of blindness at the board. All these ups and downs come out in the end in one's rating.

So, when we say a GM has a rating of 2700, that is the balance of his performance over hundreds of games. It's a very small margin for error, with the exception of very young players and a tiny handful of wildly inconsistent GMs. Chess machines aren't like this at all. When I was asked after the first match about Deep Blue's strength, estimated by the result at 2700, I said, "Yes, 2700 maybe, but 3100 in some positions and 2300 in others." In sharp, tactical play, Deep Blue could be counted on to perform far above even my level 2800+ level. This was true even of the PC engines that were still relatively weak at the time. In closed maneuvering positions, where Deep Blue's powers of calculation were muted, it could make strange and pointless moves that even a weak human master would never consider on general principles. Its evaluation ability was weak overall and, in some areas, such as a few I exploited in our match, it was terrible.

I failed to take this into account when I estimated how much it could improve in a little over a year. On a practical level, it wouldn't be decisive if the expected speed increase bumped it up another ply and another 100 points—if that 100 points went into the types of positions where it was already stronger than me. Raw speed would also have an impact on its positional play, but smaller, and if it only went up from 2300 to 2400, and if I could reach those types of positions again, I thought I would be in good shape.

Unfortunately, the Deep Blue team was very much aware of this as well. Unlike me, his former star pupil, they heeded Botvinnik's rematch rules and focused on their weaknesses. Almost from the first days of preparation, they decided they needed to put most of their efforts into improving its evaluation abilities. This meant hiring more Grandmasters to tune it and, contrary to original plans, fabricating a new set of chess chips with the new evaluation function built in. Murray Campbell and Joe Hoane wrote new software tools to make the tuning process far more efficient. The strong Spanish Grandmaster Miguel Illescas was brought in to help Joel Benjamin with the book and to play training games with the machine to further improve its evaluation. Soon, according to Hsu, Deep Blue was beating the best commercial engines even when its processing power was reduced to roughly equivalent levels, meaning it had become far smarter than before. I would be facing a very different program, not only a faster machine.

Soon after the Philadelphia match ended in February, I was invited to IBM's headquarters in Yorktown Heights for a visit, accompanied by Frederic Friedel and my new US-based agent, Owen Williams. It was a friendly affair, launching rematch talks and giving a lecture on a few moments during the match, with Deep Blue analyzing along with me. I pointed out a few weak spots in Deep Blue's analysis, which might not have been such a good idea. I was still treating it like a joint science experiment. I never would have advised Karpov on how to beat me! I spoke via remote to a couple of IBM labs overseas, including one in China. It felt like the beginning of a partnership, and I hoped it would become one. A few months later, we agreed on the basic framework and timing of the rematch: it would be in New York City in early May 1997, again over six games. Negotiations would continue throughout the year, eventually settling the prize fund and other details. The purse would more than double, to $1,100,000, with $700,000 going to the winner.

This more conservative prize fund split has been used to argue that I was less confident this time around. After all, I had proposed winner-take-all for the first match before settling for a 4–1 split of the $500,000 purse. That may well be true, although I don't recall thinking that way. And it's not as if the money was the biggest factor. I could have made more than that with far less exertion by playing exhibitions. With such a large purse for such a short match, it just made sense to hedge my bets. Guaranteeing that I would get as much for losing as I did for winning the first match was a good insurance policy. I was confident, but I knew that anything can happen in just six games. I could also be a slow starter in match play. In my five world championship matches against Karpov, I was ahead after six games in only one, our last match in 1990. In the other four I was behind in three and even in one after six games, but didn't lose any of them in the end, winning two and drawing one. (Our first match was terminated after I came back from 0–5 to 3–5.)

The rest of 1996 was a busy year for me personally and professionally. Many changes were in the air, and dealing with rematch negotiations and preparation were far from my top priorities. It was getting hard to tell if all my other dealings were distracting from chess, or if chess was distracting from them. Owen was trying to leverage the match into part of a larger project with IBM for a series of chess events, a website, and more. With Intel's departure from the PCA I was scrambling to find new sponsors, landing Credit Suisse for a Grand Prix event in Geneva in August. A month later I led the Russian team to a gold medal at the Chess Olympiad in Yerevan, Armenia. At the end of the year, I won one of the strongest tournaments in history in Las Palmas, undefeated with all my top rivals present. But my biggest "win" that year was the birth of my son, Vadim, in October.

OUR CONTACTS with IBM in the run-up to the match revealed one last flaw in my estimation of my chances. Gone was the friendly and open attitude that had been on display around the Philadelphia match run by ACM. With IBM in charge from top to bottom, this chumminess had been replaced by a policy of obstruction and even hostility. Had I paid closer attention to the media and statements coming out of IBM during the interim year, perhaps I would not have been so surprised. In August, Deep Blue project manager C. J. Tan had told the *New York Times* quite bluntly that "we're not conducting a scientific experiment anymore. This time, we're just going to play chess."

I'm no shrinking violet myself, of course. I had a thousand battles behind me and was at home in the worlds of political maneuvering and psychological warfare. My early battles with Karpov had pitted me against a Soviet Grandmaster chess player at the board and a pack of Soviet Grandmaster bureaucrats in the boardroom. Had I known going into New York that the tune had gone from a Chopin waltz to a Tchaikovsky march I would have had no trouble adjusting my own demeanor accordingly. This was difficult to do at the time, however, especially since IBM was not only my opponent, but also the host, organizer, and sponsor of the match. I had hoped they would be even more, a partner.

This gets back to the biggest reason I agreed to a prize fund that was less than everyone thought I could demand (especially my agent): I believed IBM's promises of future collaboration. During my visit to their offices in 1996, I met with a senior vice president who assured me that IBM would step in as a sponsor to revive the Grand Prix circuit of the Professional Chess Association. We had other big plans for cooperation as well—for a big web portal, exhibitions, all sorts of ways to promote chess and, of course, IBM technology. They even sent a team to Moscow to meet with me and a few of my friends to discuss the launch of the Club Kasparov website. I had no reason to doubt IBM's commitment to these grand plans until the day the contract arrived and there was no mention of any of it at all. We were told that the advertising department in charge of the budget hadn't approved it, sorry, let's play chess. That was my first notice that the gloves were coming off for the rematch. C. J. Tan and others still occasionally referred to future cooperation with me in public during the rematch, but it was only for show.

It was a disappointment because I had invested time and resources in what I thought was going to be a great coup for chess. It also marked the first feeling of betrayal of the experiment I thought I had joined when I played Deep Thought in 1989, the longest-running science experiment in history. I'd met the team and been impressed by their dedication and ambition. There was mutual respect then, and at the first match in Philadelphia. By the time the rematch neared, it was clear IBM didn't want my respect or my partnership; they wanted my scalp.

As they never ceased to remind me, I had agreed to the rules long ago and couldn't complain later if they exploited them to the letter. A case in point was my request for all the games Deep Blue had played in the previous year. Prior to the first match, these had been made available freely, although there weren't very many. Before the rematch, my request was met with a terse reply: there weren't any games and none would be forthcoming. We knew that Benjamin and Illescas and others had played training matches against Deep Blue, although they had purposefully withheld the machine from public competitions for the entire year. In fact, it turned out later that we vastly underestimated how much other GMs participated in the project. The PCA might have been falling apart, but thanks to me, IBM had become a source of employment for quite a few Grandmasters. We were told that since those were not official games, as specified in the match rules, they were under no obligation to share them with me. No games.

When I brought this up at a prematch press conference, Tan's reply was that I would have to send them all the training games I had played against other computers. I had played dozens of tournament games in the past year that they had ready access to, but I immediately answered that I would be glad to hand over all my training games against the engines Fritz and HIARCS. But IBM never responded to this offer and so Deep Blue would be a black box until game one. Another concession that came back to haunt me was the schedule. I knew I would need all the rest I could get against an opponent that needed none, especially since the Philadelphia match had taught me how tiring it was to play classical games against a machine. Instead of insisting on a rest day before the final round, I foolishly agreed to two consecutive rest days after game four, so that games five and six could be held over the weekend, possibly improving attendance and coverage. It was a mistake that would have enormous consequences.

That press conference was my second notice that the experiment was over and that the friendly competition was done. No more shared meals and chit-chat about the games like during the first match. My presumption of continued good faith was exposed as naïve. It was a rude awakening. When I was asked what would happen if I lost the match, I answered, "Then we'll have to hold another one under fair conditions." Rude, I suppose, but it was only now that I could see the direction things were going. I was annoyed at myself for being so easy-going when the rules and other arrangements had been made. After the first match, it simply didn't occur to me that things would change so much. I could only hope that this new secretiveness and antagonism didn't extend to affecting the match in any way.

This was another flawed assumption, because IBM had performed a simple equation when they decided to go all-in to win. Despite the Deep Blue team's tremendous efforts, it wasn't clear to them that they would be able to get the machine up to my 2820 level. And by the time the match started, even with the new evaluation-tweaking tools and the opening book, they couldn't make Deep Blue play any better. But there was always the chance that I could be induced to play worse. Deep Blue didn't have to play at a 2800 level to beat me if I didn't play at that level myself. And so began the games within the games.

CHAPTER 9

THE BOARD IS IN FLAMES!

IBM HAD TAKEN OVER several floors of the Equitable Center in midtown Manhattan for the match. Deep Blue's main system would be on-site, in a room with more protection than any in the Pentagon. According to Newborn, there were several backup systems connected to it, one in Yorktown Heights and a smaller one in the building that could take over seamlessly. The new Deep Blue was operating on a new supercomputer model that was twice as fast as the old one, and it contained even more of Hsu's new and improved chess chips, 480 of them, and reached 200 million positions per second at its peak. I read much later that this new version beat the old one at a three-to-one ratio in training matches, but that wouldn't have meant much to me had I heard that before the match. Even a largely unimproved version of the same program will be much stronger than its old self at double the speed; there is no easy way to translate how well a machine does against other machines to how well it does against Grandmasters.

The playing area was a small room with a VIP seating section with around fifteen chairs. On another floor, there was a large auditorium that seated five hundred people equipped with large video screens so they could watch us at the board while following the live commentary. US Grandmasters Yasser Seirawan and Maurice Ashley did most of the analysis, alongside computer chess expert and IM Mike Valvo. Another addition to the commentary team was Fritz 4, and I suppose it was only fair to have one of the hosts giving the machine's point of view! The audience was very much on my side, as you would expect, which was always a little awkward for the IBM team. It was their event, top to bottom, but their guests were always rooting against them. The good news for them was that their player could not have cared less about home-court advantage or fan support.

A few days before the match started, we had an inspection of the playing area and the facilities my team would be using during the games. My assigned rest area was quite a walk away from the playing room, so that needed to be changed and it was. This area is mostly for pacing and a quick drink or snack during the game. Deep Blue needed thousands of watts of power to play chess; the twenty watts my brain used during a game only needed bananas and chocolate. This fact stems from one of the more intriguing ideas I later heard about adjusting the playing field in human-machine competitions: energy equality. That is, a chess machine that doesn't use any more power than the human would represent an enormous advance in energy efficiency.

The next surprise came when we asked about where my team would be during the games and we were told that, contrary to IBM's conversations with Owen, we had no team room. They would have to sit in the press room or with the audience, rotating with my mother between the two seats allocated. It was all very strange to feel like an afterthought to the organizers. Even simple requests often went through multiple channels and delays. I admit that I was accustomed to first-class treatment at chess events. Like Bobby Fischer before me, as world champion I believed it was not only my right to be demanding about the conditions, but my duty, since it set a standard for other events and for other players. A couple of small slights and hassles could be nothing, but when they form a pattern it is cause for concern.

Before the games get under way, I will take pains to note that few or none of these concerns and grievances about the atmosphere and organization before and during the match are intended to reflect negatively on Deep Blue's creators. Inevitably, since they were participants and IBM employees at the same time, they were thrust into an adversarial position when I made demands or filed protests. I have already said that I'm not convinced that even a world champion machine's programmers and trainers have earned the arrogance of a human world champion, but they were fierce competitors and I cannot begrudge them for it. C. J. Tan mostly handled these things, but the Deep Blue team could hardly help being drawn into the mêlée during press conferences and interviews. I was a veteran of seven world championship matches and I knew that I had to push back against the increasingly antagonistic match organization or I would feel psychologically crushed. Inexperienced in such things, and put on the firing line by IBM PR, Campbell, Hoane, and Hsu felt like I was being hostile to them, and perhaps sometimes I was. This was another example of the problems of having IBM as both the organizer and a participant.

GAME ONE of the rematch might have been the most anticipated game of chess since the first game of the Fischer-Spassky match in 1972. Magazine covers, bus-stop ads, TV talk shows, you couldn't miss it. The press room at the venue overflowed and had to be moved to a larger space. I tried to enjoy it, and then to ignore it, but the pressure was already mounting. Yuri, Michael, Frederic, and I worked out a general match strategy that I hoped would allow me to learn as much as I could about this new Deeper Blue without taking huge risks. My world championship matches had lasted weeks, even months. Over the course of sixteen or twenty-four games you had time to experiment, to try different ideas. In only six games there would be no time to recover from an unforced error.

Months before the match, I said in an interview that "the first match proved that in certain positions the machine is unbeatable, and in certain positions it is hopeless. Certainly, there are many positions in between. I know in general what to expect, but I am cautious about surprises."

In New York, I'd been listening to the IBM team all week talk about how they had improved Deep Blue dramatically, and bumping into several American Grandmasters who had been working with them was a bit of a shock. My team found out during the third game that, despite IBM's statements that they hadn't been working with other GMs, several were staying at the hotel with the rest of the IBM team. The *New York Times* reporter would later confirm that they had been hired by IBM.

As with all the other little surprises, it was another indication I was in for an all-out battle. When preparing for a big match, one's seconds are usually a closely guarded secret. If you know with whom your opponent has been training, you might be able to divine which openings they are preparing. If you are planning on playing the Sicilian, for example, it would make sense to hire an expert in that defense. Had I seen the Deep Blue team hanging out with a few of the world's top computer scientists I'd have assumed they were finding ways to increase its speed and not worried so much. As I said, Deep Blue going from 3100 to 3200 in tactical positions wouldn't be decisive if I could avoid those positions. But if they had been working with a big team of Grandmasters, maybe they were really teaching it to play chess! Raising its positional evaluation level up to the GM range of 2500 would render it immune to most anti-computer ploys.

And with such a large team, it was a sure thing that they had spent a lot of time on the opening book. When I played a world championship match, my team and I had been trying to outprepare my opponent and his team for months. But when we walked out on stage, it was just the two of us and our memories doing battle in the opening. Deep Blue didn't have to worry about forgetting any of the thousands of opening lines it had been fed by its Grandmaster tutors.

This was just one of the many complicated asymmetries in human-machine chess, and there wasn't much to be done about it once the rules had been agreed. Later such encounters, learning from my experience in New York, would include stricter regulations to attempt to level the playing field. For example, limiting how many variations could be added or altered to the machine's opening book between games and providing the human player with a relatively recent version of the engine a short time in advance to compensate somewhat for the lack of published games. (The rules of play for the Deep Blue rematch were three pages long. For my next human-machine match there would be over six pages. Humans can learn, too.)

Other regulations would address even thornier issues of fair play and secrecy, which were also an asymmetrical problem with no perfect solution. For example, if the machine crashed or had some other problem during the game, should the human player be informed? That would disturb him, but otherwise your mind could run wild wondering why the operator had suddenly started typing, or hurrying back and forth to discuss things with other team members. Another was that there should be a detailed log of all the human interactions with the computer during the game, not only those of the operator. Remember in Hong Kong, when people back in New York were working on getting Deep Blue running again during its match with Fritz? Remote access and redundant backups make monitoring the machine's activities a nearly impossible task, requiring both technical expertise and total access at multiple sites. In my team's preparations for New York, we simply weren't alert enough to the potential for distraction caused by worrying about such things, again having been unwise to believe the match would be as open and friendly as everything had been in Philadelphia.

The need for such rigorous oversight is to provide peace of mind by guaranteeing that any disputes that do arise will be handled equitably, without giving advantage to either side. This is critically important when only one of the participants has a mind that requires peace to perform at top level. If there is an atmosphere of good faith and openness, the fine print is not so critical. There might still be accidents and issues that aren't covered in the rules, as happened in several of my other machine matches. Sometimes there is just no way to avoid one side being punished for something beyond their control. Do you forfeit a machine after multiple power outages in the building? That is unfair to it, clearly, but what about the human competitor, distracted and tired, sitting in the dark wondering if the game will continue? How long do you wait?

A few days before the match began, I was sitting in a more metaphorical darkness, one regarding Deep Blue's capabilities. I was bitter over having been stonewalled about seeing any of its games. What was I supposed to base my preparation on? I knew that the six games I had played against it in Philadelphia were too small a sample size to be reliable, especially since they would have worked to solve precisely the problems I exposed. I decided I would try to use the first few games to see if I could get a sense of its strength and tendencies. This meant playing more passively than I preferred, although it fit my general strategy of wanting to play quiet positions where its tactical abilities would not be the deciding factor.

Predictions for the match were largely in my favor, outside of the IBM team, of course. Some, like David Levy and Yasser Seirawan, even thought I would surpass the 4–2 score of the first match, since I would be able to build on that experience. For my part, I was as bold as ever in my predictions. Why not? Has any sportsman ever gone into an event predicting the futility of his participation? But I was genuinely confident, based on the reasons above about how much I thought they would be able to improve Deep Blue's play in a little over a year. IBM's Tan surpassed even my bravado, saying Deep Blue would win the match "overwhelmingly."

The drawing of lots took place at the Equitable Center on May 1. This is an old chess tradition to determine who will have white in the first game, and it's usually embraced by the organizers as a way of injecting a little local color into the proceedings. When there are no props, who plays with which color is decided simply by one player holding pawns of each color behind his back while the other player chooses a hand. If he picks the hand with the white pawn, he gets white. Too boring. Over the years, I have participated in some of the strangest methods of deciding colors you could imagine. There have been lottery balls, animals, dancers, and magicians. At a 1989 tournament in Skellefteå, Sweden, the players were confronted with sixteen real gold bars. Numbers were taped to the bottom of each bar, and all sixteen players stepped forward to lift one to find out their starting position. Seeing some of the others struggle with the weight, I steeled myself to try to lift mine with only one hand. But I failed and had to use both hands like the others, only to watch as Hungarian GM Lajos Portisch, twice my age, lift his one-handed with no sign of exertion. In 2002, I played a rapid match in New York's Times Square against Karpov. The amazing magician and chess aficionado David Blaine was in charge of the drawing of the lots and it appeared he would do it the old-fashioned way, with two pawns. But of course, the pawns kept disappearing and disintegrating in his hands!

Things were a little more sedate in New York as C. J. Tan and I were presented with two identical boxes containing New York Yankees baseball caps, one white and one black. I chose a box and inside it was the white hat—fitting for a defender of humanity! In a reversal from the first match, I would begin the rematch with the white pieces. This was not inconsequential, at least in theory, because under the circumstances I would have preferred to finish with two whites in the last three games in order to better exploit what I had learned in the first games. As demonstrated in the first match, having white in the final game can also be a tactical advantage, depending on the match score. If your opponent is even or behind, he, or it, is under tremendous pressure to capitalize on his last turn with the white pieces. Also, I was effectively conceding a good part of the advantage of the white pieces with my cautious openings at the start, so starting with black would have been my preference.

THE DAY FINALLY ARRIVED. Hundreds of journalists were there to cover it live and the auditorium was full. I shook hands with Hsu at the board and tried to put all the distractions out of my mind as an army of photographers fired away. It was a relief to finally play chess, and to finally see what this thing was made of. I was happy to note that the weight of humanity's defense did not make my pieces noticeably heavier.

I opened game one by moving my king's knight to f3, just as I had in all my white games in the first match. It's a flexible move, allowing many transpositions by both sides, ideal for sounding out my opponent. It was part of my "anti-computer" strategy, as unhappy as I was then to do it and as unhappy as I am now to write it. I would have loved to play the same sharp openings I usually played against the likes of Karpov and Anand, to match my preparation against a computer accessing a library as infinite as that of Borges.

But I also had to be practical. I wanted to win, not go out in a blaze of glory, no matter how honorable the flames. I knew from practicing against far weaker programs that the razor-sharp positions I preferred against any human in the world could be trouble against Deep Blue. I was confident I would be fine in the opening—I would pit my preparation to this day against any team of Grandmasters in the world. But there were two big problems with following my main lines against Deep Blue.

First, being able to simply regurgitate moves from its opening database and perfect recall of my own games would give the machine a free pass to the middlegame where it excelled. Why let the machine get to move twenty playing as well as Karpov because it was literally repeating the moves of Karpov? Frederic had shown me commercial opening books that were so deep in some variations that they ended practically at the endgame. If Deep Blue really could play at the world championship level, I wanted it to prove it by thinking, not by pantomiming my own games back at me. I hoped to exploit its inability to plan or to play strategically by getting it out of its book as early as possible, even if the position was not objectively great for me. At the very least, even if we transposed back into main lines later, as happened several times, I would gain some insight into its preferences in the process.

Second, many of my favorite openings led to sharp, open positions where we would be closer to the territory where Deep Blue played at a 3000 level and further from the closed, maneuvering positions where it played much worse. Even if they had improved its positional understanding as much as they said, and this indeed seemed to be the case, I thought my chances would be better in an anti-computer bog than in a pitched battle on the open plains. It was quite painful to make this choice. I am by nature uncompromising, at the chessboard and elsewhere. But I cannot say it was the wrong decision just because I lost the match.

One of the mistakes of "narrative" in game analysis is what we call "analyzing to the result." That is, the winner made good moves because he won; the loser blundered because he lost, etc. Since you know the outcome of a game before you begin to analyze it, it is very difficult not to eye the eventual loser's moves more critically, even when it may not be merited. Knowing I lost the rematch to Deep Blue makes it easy to view all my decisions as mistakes, when each should be evaluated as objectively as possible. Losing a game or a match does mean you made mistakes, of course, but we should also remember that, as the American chess author I. A. Horowitz wrote, "One bad move nullifies forty good ones."

My anti-computer strategy did pay dividends in the first game, if not conclusive ones. I proceeded with the Reti Opening I had used against it before with success, and eventually we transposed to a well-known position where I could be sure that Deep Blue was still in its opening book. I then deviated on move ten with a move that I would be embarrassed to play against a human opponent. Instead of the normal expansion in the center by pushing my king's pawn two squares, I advanced it timidly a single square, avoiding contact with black's forces. It was intentionally passive, almost a waiting move, a throwback to David Levy's old trick of seeing if the computer could be fooled into weakening its own position when left without concrete targets.

And lo and behold, it could! It's next move unnecessarily created a weakness around its king. Instead of exploiting my tame play, Deep Blue didn't know what to do with the extra time I had given it. Remember that these were the first dozen moves anyone outside of the IBM camp had ever seen this Deep Blue play. For me this was a good sign that it still had something to learn. Now the question was whether I could teach it. Playing possum might encourage a few mediocre moves from it, but I knew that if I was going to win I would have to go on offense at some point.

I continued my maneuvers and was rewarded again by two pointless moves by Deep Blue. I read later that the GM commentators and the audience laughed at the machine's shuffling. Such time wasting didn't really put it in danger in such a quiet position, but it gave me confidence and it gave me an idea. I made a threatening move with my knight, hoping to encourage Deep Blue to make another weakening pawn move in front of its king in order to preserve its bishop. To my gratification, it obliged, forcing my knight back but leaving its position full of holes that I could target later.

It would not be easy, however. Computers could often be led into creating weaknesses in their positions, but they were also incredibly good at protecting those weaknesses. There is no value in a theoretical weakness; you have to be able to exploit it. Deep Blue was making strange, inhuman moves, but they weren't necessarily bad moves for a machine. It wouldn't matter very much if the objective evaluation was in my favor if it reached the sort of position it could play well.

This is not to say that my position was so good objectively. I had played so cautiously that I was not in position to take advantage of Deep Blue's weakening moves. But this was according to my overall match plan. I had to keep reminding myself not to rush, that I needed to find out as much as I could about my opponent's abilities. My priority was to limit the machine's counterplay, as I had in our last game, when I squeezed the life out of it in game six in Philadelphia. But this version of Deep Blue was much improved and was not going to allow itself to be squeezed, which meant eventually the gloves would have to come off.

English Grandmaster John Nunn writes about this moment in his ChessBase analysis of game one: "This is the critical phase. Everybody who has played a computer knows the scenario: you get a strategically winning position, the computer makes some desperate tactical lunge, you make a couple of inaccuracies and suddenly the machine is all over you." Indeed, Deep Blue found some very strong moves to counterattack before I could consolidate further. It launched its pawns forward, probably the first time a computer attack has ever made an audience gasp. It was stirring up just the sort of slugfest I had attempted to avoid. The time for caution and prophylaxis was over. It was time to meet fire with fire, or, as match commentator Ashley said to the audience at the time, "The board is in flames!"

In a decision that commentators at the time and in the many articles and books on the match have called "bold" and "crazy," I allowed Deep Blue and its two raking bishops to rip open the position around my king. I was counting on an exchange sacrifice—giving up a rook for a bishop—and the power of two pawns pressing in on black's king. As GM Danny King writes in his book, *Kasparov v Deeper Blue,* on the match, "Both man and machine must have arrived at this position in their calculations several moves earlier, and both must have judged it to be favorable for themselves. It's a close one."

As Prussian field marshal Helmuth von Moltke said, no battle plan survives the first contact with the enemy. My plan for a quiet fact-finding mission in game one had been blown to hell by the aggressive machine. I was pinning my hopes on my superior evaluation ability. Deep Blue liked its material advantage and well-placed pieces. I liked my two connected passed pawns and powerful dark-squared bishop. It was a classic duel of imbalances in a dynamically equal position. It was a mêlée, but I had enough time on my clock that I was confident in my ability to deal with whatever tactics arose.

After I beat him heavily in a match in 1986, English GM Tony Miles called me "a monster with a thousand eyes who sees all." I didn't like that nickname any more than I liked being called the "Beast of Baku" (*el Ogro de Baku* in Spanish, I'm told), but I suppose it was a compliment. My ability to see in seconds what even experienced Grandmasters needed minutes to work out was what first drew me to the attention of Mikhail Botvinnik when I was a child. I wasn't a machine, or an all-seeing monster, but I was about as close as a human could get when it came to chess. GM Robert Byrne wrote in the *New York Times* the next day in an article titled "In Late Flourish, a Human Outcalculates a Calculator": "In overcoming the marvelous IBM chess computer Deep Blue yesterday, Garry Kasparov beat it at its own game."

Had Deep Blue realized the position was roughly equal it probably would have been fine. Instead, it overestimated its material advantage and happily exchanged queens when it shouldn't have. It was a classic computer mistake: it was happy with the status quo but couldn't see that it would have no ability to improve its position, while I could and did. Deep Blue had one last good chance to escape with a draw. To do so, it would have to give back its material advantage, but it turned out even a machine can be too stubborn for its own good. Instead of doing the equivalent of admitting its mistake and bailing out, Deep Blue tried to hold on and went down with the ship. After another defensive inaccuracy and one very strange rook move I'll discuss in a moment, black's position became hopeless and Campbell reached out his hand to resign. Remarkably, none of my pieces had ever crossed out of my half of the board, a very rare occurrence in a victory. My pawns had, however, and that turned out to be enough.

I arrived at the auditorium, where a standing ovation welcomed me and then also the Deep Blue team. We both deserved it. It had been a real battle, a rich game of chess. I had emerged with the victory, but as I said on the stage after the game, it already felt very different from Philadelphia. This Deep Blue was a worthy opponent.

I HAD LESS THAN twenty-four hours to savor my third consecutive victory over Deep Blue. I would have black in game two and needed to be well prepared. Having the advantage of moving first doesn't mean very much in amateur play. The extra tempo—a beat in time on the board—was worth less than half a pawn at the start of the game. This is practically insignificant when weak players are exchanging blunders and wasting time on nearly every move. For Grandmasters, every tempo is precious, especially in sharp positions where the victory will go to whomever's attack lands first.

In relatively closed positions, such as the early part of game one, losing a few tempi wasn't fatal, if also not desirable. Levy's old anti-computer maxim was to "do nothing but do it well" and let the machines throw themselves to their doom against his well-prepared defenses. In a sort of reversal, Deep Blue didn't know what to do against my temporizing, but it did it well enough not to get into serious trouble. When the smallest opportunity arose, Deep Blue struck quickly and struck hard. It's not accurate to say that I wouldn't underestimate it again, because I had no information with which to estimate it in the first place. But going forward, I wouldn't underestimate it and I wouldn't give it a free swing at my chin with the white pieces in game two.

As night follows day, it wouldn't be a victory over a computer without a bug report. Bugs to chess programmers always sound like the way GMs often say we "forgot" something during a game instead of admitting we simply missed it and our opponent didn't. Spassky joked about this tendency in an interview in 1988, saying about a book of his games he was working on, "I want to be very honest. If I didn't see something I would like to say, 'Here I was blind, I didn't see this!'" Perhaps Shakespeare said it best when he wondered if a mistake by any other name was still a bug.

Of the two "bugs" mentioned from the first game, only one has been discussed as consequential, and not for its impact on the game in which it occurred. And in a strange twist that is another illustration of the power of narrative over fact, this bug got a second life fifteen years later.

By move forty-four of game one, the game was essentially over. My position was winning and we had passed move forty, so I had plenty of time on my clock to avoid any tricks or accidents on the way to victory. Modern engines evaluate the position after my forty-fourth move at nearly +12 for white, more than the value of an entire extra queen. A human in such a position would slump in his seat in despair, thinking dark thoughts about the earlier mistakes that had brought about his ruin. Computers don't do that; they keep churning through billions of positions looking for the best move. They don't understand the human heuristic of practical chances, that when you are in deep trouble anyway, it's often better to play an objectively inferior move that might confuse your opponent. Nor did machines have pride to interfere with their calculations. To a computer, a groveling move that allowed checkmate in ten moves was still clearly better than a tricky move that allowed checkmate in nine. Anyone with experience facing computers knew that when they were facing imminent death, they could make bizarre-looking moves to postpone mate just a bit longer.

Deep Blue's forty-fourth move looked like one of those moments. My pawns were close to promoting to queens and there was no way to stop them for long. If I could see it that clearly, I knew Deep Blue could as well. Maybe it had worked the position all the way out to checkmate already, not unrealistic in a forcing position with few alternatives—a very narrow search tree. Instead of resigning or making one of the defensive moves I was analyzing, Deep Blue played its rook down the board, away from the action. I couldn't imagine the point of it at all, so of course I had to triple-check to make sure there wasn't some brilliant computer geometry hidden in the move. Finding nothing amiss after a five-minute think, I happily discarded it as one of those inexplicable moves computers often make when they are totally lost and I pushed my pawn to g7, one step away from promoting to a queen. I put my Audemars Piguet back on my wrist, part of my ritual when I knew the game was ending. Campbell resigned, confirming my conclusion that Deep Blue's odd final move had been the last gasp of a landed fish.

My team and I still had to analyze the game that evening, especially the opening. We did pause when we arrived at Deep Blue's strange forty-fourth move, however, because we couldn't get our computer engines to duplicate it or explain it the way we expected. Deep Blue's move simply looked inferior, although our relatively primitive machines took a long time to work it out all the way to checkmate the way an engine can do now in a matter of seconds. (Mine says the final position is checkmate in nineteen moves, although it's trivially winning after just five.) Was Deep Blue seeing so much deeper than us and our PC engines that this move somehow made sense to it? How to explain it? "How can a computer commit suicide like that?" I asked Frederic. After toying around with Fritz for a little while I found the forced win after the move I had expected from Deep Blue, giving check with the rook. It was a pretty sequence that I hadn't seen at the board, but, I assumed, Deep Blue had. I concluded that the machine, seeing mate coming, played something that made perfect sense to it to delay the inevitable. Case closed. Computers often made unfathomable moves in totally lost positions; if we needed to analyze more moves like that, it would be very good news indeed.

Other commentators agreed with my conclusion. King's match book calls the forty-fourth move "curious" and "odd," adding that the machine probably "saw a quicker win" after the expected moves. The position was clearly lost already, so it wasn't even worth adding the usual "?" annotation to the move we use to indicate an error.

Yuri and I got back to preparing the opening for game two. Frederic, meanwhile, filed away this insignificant moment from the first game and, storyteller that he is, turned it into the stuff of legends. In his write-up for ChessBase, he dramatized my bafflement around move forty-four, despite the fact we had reached a satisfactory conclusion in our analysis (although that conclusion would turn out to be incorrect). Frederic wrote, "The conclusion was a little bit scary. . . . Deep Blue had actually worked it all out, down to the very end and simply chosen the least obnoxious losing line. 'It probably saw mates in 20 and more,' said Garry, thankful that he had been on the right side of these awesome calculations."

Harmless enough, especially since he also included my analysis with Fritz showing the checkmates I would have delivered had Deep Blue played the expected move. And the words "scary" and "awesome" are Frederic's, not mine. Somehow, after the match this little anecdote was transformed into the urban legend that I had been so impressed by the apparent depth of the machine's calculations with that strange rook move that it affected my play and my decisions for the rest of the match, especially in the critical second game. This hypothesis was proposed by Murray Campbell at least as early as Monty Newborn's 2002 book on Deep Blue. The punch line to his theory was that Deep Blue's mysterious move wasn't profound at all; it was a blunder and the result of a yet another bug. Per Campbell and Hsu, the move was "random," the result of a known bug they had failed to kill before the match began.

This tale acquired new life when election analyst Nate Silver used it as the centerpiece for an entire chapter of his 2012 book, *The Signal and the Noise*. The narrative suggested by Frederic and spread by Campbell was irresistible: Kasparov lost to Deep Blue because of a bug! Writes Silver, "The bug was anything but unfortunate for Deep Blue: it was likely what allowed the computer to beat Kasparov." *TIME*, *Wired*, and other outlets ran with breathless variations on this theme, each story containing more errors about chess and more silly assumptions about my mental state than the last.

I'm sincerely glad that my matches and chess have acquired the cultural cachet to become the subject of so much popular writing and other appearances in pop culture. The problem is, that just like most of the chessboards you see in the movies are turned sideways, the people writing about chess in popular publications often have absolutely no idea what they are talking about. Instead of taking the time to consult with a professional player, they take it for granted that winning a plastic trophy in a second-grade chess tournament qualifies them to comment insightfully on the moves and mindset of a world champion.

Much of what Silver gets right in his chess chapter is from other sources and it stands out when so much else is wrong. Among many other things, he misunderstands the workings of opening books, calls the middlegame "the midgame," and makes a complete hash of game six. That is still to come, of course, but for one sample he writes, "Kasparov did not know the Caro-Kann [Defense] . . ." It's true I had given up playing the Caro-Kann in my youth, but I also cowrote a book on it. It's also obvious to any chess player that you can know an opening quite well even if you don't play it yourself if it's one that you regularly have to play against, as I did.

Getting back to the game at hand, Silver ignores the fact that Frederic writes in the very ChessBase article that started it all that "somewhere around here Fritz started to announce mates." Even little Fritz on a home computer was seeing well over a dozen moves deep, and we knew Deep Blue was much, much faster. It was well understood that searching that far into a position was quite possible for a machine if the moves were mostly forced and material was limited. The search tree narrows dramatically and "singular extensions" such as those Deep Thought had a decade earlier can push very far. When there are checks against the king and only four rooks and a few pawns on the board, as was the case at that point in the game, Deep Blue easily could have reached that search depth in a few minutes. You can even see in the machine's log files, published years later, that it reached a depth of twenty ply a few moves earlier, on move forty-one when there were even more pieces on the board.

Had Deep Blue done something as mysterious in an equal position, that would have been another story, one requiring investigation. Coming as it did at the end of the game in a totally lost position it was curious and quickly forgotten. I was at first confused, and then slightly impressed, nothing more. But the allure of "a bug beat

Kasparov" was too strong even for a statistician. Turning it into amateurish psychoanalysis and the reason I resigned in game two is absurd mythmaking.

Silver begins by citing Edgar Allan Poe's 1836 essay on the chess-playing hoax automaton the Turk, but he should have taken into better account something else of Poe's, "Believe nothing you hear, and only one half that you see."

I will agree with the broader conclusion of this little story, that I would not have lost the match had I been in a better state of mind. But that state would not suffer until game two and its incredible aftermath.

IF ANYTHING, my confidence was very high going into the second game. I had beaten Deep Blue three games in row going back to Philadelphia and it was a huge relief to no longer be playing a ghost. This Deep Blue II was strong, but it was far from perfect. It had made a series of computerlike inaccuracies in the opening phase, although it recovered well. I had taken it on tactically, proved my evaluation of the position was superior, and come out with the full point.

Game two would be another matter because I had the black pieces. After seeing how aggressively the machine played when given the chance, we decided that using the same passive anti-computer strategy with black would be too dangerous. With white, I could control the pace of the game much better and wait for my chances. With black it would be safer to play a known opening even if it was in Deep Blue's book, especially if it was a closed opening where it would have difficulty finding a plan. The downside with this strategy, as in all the games, was that it wasn't my style either. While I was playing anti-computer chess, I was also playing anti-Kasparov chess.

Whether that was the correct strategy isn't any easier to decide in hindsight. Had I been in possession of even a dozen of Deep Blue's games to get a sense of its capabilities I would have felt comfortable playing my usual openings and preparing for it like I would any Grandmaster opponent. Without anything on which to base concrete preparation, it felt best to stick with flexible positions where I didn't have to add worrying about opening novelties to the long list of things I had to worry about. Energy conservation was a major factor in my off-board calculations. Playing against a machine was exhausting because I was obliged to look at possibilities I wouldn't normally consider, and to double-check every calculation. During a regular tournament or match, Yuri and I would be up late the night before each game, trying to squeeze in every second of prep time for the next day's opponent. That would be a disaster against a machine that wasn't going to tire at the same rate as I was.

One thing I am sure of, my opening choice in the second game was the worst of both worlds. It was a Spanish Game, also called the Ruy Lopez, named for the sixteenth-century Spanish priest who analyzed it in one of Europe's first important chess books. Its nickname is the "Spanish Torture," and the reason why will soon be painfully clear. I didn't want anti-computer chess and I also wanted to avoid my usual sharp Sicilians in order to avoid walking into a surprise in Deep Blue's book. The Ruy Lopez is generally a quiet maneuvering opening, one of the most strategically complex and deeply analyzed systems in the vast halls of chess literature. Many of its main lines have been investigated well past move thirty, the point at which many games end.

It was not one of my openings with black, although I did have extensive experience trying to beat it with the white pieces. The Ruy Lopez had been one of the central battlegrounds of my world championship matches with Karpov, Short, and Anand. I wasn't happy about giving Deep Blue a database free ride into the middlegame, but we decided it was worth a try. Deep Blue hadn't demonstrated any improved ability to play positionally in the first game. I hoped to be able to keep the position closed, leaving it without a clear plan to make progress. At that point, if things were favorable, I could try to press a little myself. Otherwise, with everything blocked, a draw with black and a lead in the match was a perfectly good result.

It is usually clear when playing against both people and computers when they are still "in book" due to how they reply instantly, without thinking at all. If you have a move memorized, and already know that's the variation you want to play, why waste time on the clock?

There are several nonrhetorical answers to that rhetorical question. Sometimes you simply want to get your bearings and double-check to make sure that you aren't wandering into any traps or tricky transpositions. Chess has been described as trying to paint a masterpiece while someone yanks at your sleeve, and both players feel the same way. You must always remember that at every moment of a chess game, the position is a joint creation. So, while you may be happy with how things are going in the opening, it's usually fair to say that your opponent is also content to be there, which should make you at least a little cautious.

The other reason you might pause before making a move in the opening is psychological gamesmanship. Most players are inclined to bash out their opening preparation at a rapid pace, which can have a good psychological impact, especially if your opponent is thinking deeply on every move. It is unsettling to be working hard at the board in a complicated position and, when you finally produce your move, your opponent replies instantly and puts you right back to work. This can also be intimidating because you know your opponent knows the position better than you do, and, in this day and age, it likely also means he prepared that position with the help of a very strong engine. In effect, you're not only playing against your human opponent, but against his computer as well. In an ironic but inevitable turn, machines use opening books based on human knowledge, but increasingly Grandmasters employ engines to assist them with their opening preparation.

But you don't always want your opponent to know that you are still in book. Perhaps you have prepared a nice novelty down the line and you don't want to arouse his suspicions by racing ahead too eagerly. A pause here and there may lead him to believe you have not prepared so deeply in the variation you are playing, giving him false confidence. I generally didn't have the patience for such deceptions. I was incredibly well prepared and wanted my opponents to know it. As Fischer once said in an interview, if a bit disingenuously, "I don't believe in psychology. I believe in good moves!"

Of course, machines are not vulnerable to psychology, although I suppose it might be useful to leave their human coaches wondering about your level of preparation. It was quite a shock, however, to find out twelve years later that although Deep Blue was immune to gamesmanship, it was very much capable of employing it itself.

BASED ON interview comments from Miguel Illescas and others, it is my understanding that everyone who worked with Deep Blue signed nondisclosure agreements forbidding them to talk about what went on behind the scenes without approval. It's hard to imagine a clearer illustration of C. J. Tan's pre-rematch statement that "the science experiment is over" than banning free discussion of how Deep Blue achieved its momentous feat. It's also hard to imagine the point, especially for the nontechnical advisors. It's not as if there was a competing chess machine project out there ready to hire Deep Blue's Grandmaster coaches for their secrets. Why didn't IBM want its team members to speak to the press? And for ten years?

The Spanish theme of game two is a good time to bring this up, because in 2009, Spanish Grandmaster Illescas broke the silence and gave a long interview to *New In Chess* magazine in which he spoke extensively about his work on Deep Blue and other events during the match. It only took a few paragraphs of the article before I understood very well why IBM had made them all swear to secrecy.

Illescas is an astute and easy-going man as well as a strong Grandmaster and trainer. He runs an important chess academy in Spain and also publishes a magazine there. Coincidentally, or maybe not, he was also a second for Vladimir Kramnik in the 2000 world championship match in which I lost my title. This makes Illescas the common denominator in the only matches I have ever lost in my life, but I bear him no ill will. Well, maybe a little.

I'll return to other, even more intriguing, sections of his interview later, but here he refers to Deep Blue's opening play in game two. "We gave Deep Blue a lot of knowledge of chess openings but we also gave it a lot of freedom to choose from the database with statistics. In the second game, in a Ruy Lopez, the machine was thinking about a move like a4. A very theoretical move and Kasparov was perhaps surprised when the machine started to think about a theoretical move. It thinks for ten minutes and finally plays a4. What's going on? Then he probably started to draw too many conclusions. This was a new approach for that time and Garry was never sure whether the computer was playing theory or thinking for itself."

Interesting, although I learned about this technique not long after the match. It definitely makes sense to let such a strong machine have more input in its opening choices if it was truly stronger than many of the Grandmasters whose games it would otherwise follow blindly. But what Illescas said next was a shock. "Of course we also built in some tricks for Garry. For certain moves there was a delay or some moves it played immediately. In some positions we bet that Garry would play the best move, and if he does, let's reply immediately. This has a psychological impact, as the machine becomes unpredictable, which was our main goal."

Amazing! They created programmed delays to trick me—and only me, since Deep Blue never had another opponent in its entire brief existence. It was also a one-way street, since Deep Blue was as immune to such tricks as it would be to Ruy Lopez's suggestion to always sit where the sun would be in your opponent's eyes. All's fair in chess and war, I suppose, but this revelation was additional confirmation that winning wasn't everything for IBM; it was the only thing.

AS GAME TWO is one of the most heavily scrutinized chess games in history, I'll spare you most of the torture and move directly to why it is so famous. After twenty moves of typical Ruy Lopez maneuvering, it was fair to say that both players were happy. To clarify, I did not like my position at all, but it was the sort of closed, strategic position I had intended to reach. My pieces were cramped behind the clashing walls of pawns, a typical disadvantage in the Ruy Lopez with black. White had more space, meaning more freedom to move its pieces and to probe for weaknesses. I was betting that Deep Blue would not have the patience or skill to do that sort of subtle probing.

I was the first to make a mistake, although again it was the sort of mistake I knew I was making. Following my belief that the position should be kept as closed as possible, I partially sealed the pawn structure on the queenside, leaving me with no active way to counter white's plans. It would have been a terrible position against a Grandmaster, but based on the first game, Deep Blue hadn't improved its evaluation functions enough since the first match to exploit its advantages. Slowly, however, it became clear that the Deep Blue of game two was very different from the Deep Blue of game one. It maneuvered behind the lines expertly, preparing for an eventual breakthrough. There was none of the purposeless shuffling it had done in the first game. Meanwhile, I was the one with nothing better to do than shuffle around. The Spanish Torture was under way, and I was on the rack.

Deep Blue then surprised the commentators by opening lines on the kingside by pushing its f-pawn two squares. It looked like a surprisingly humanlike move, following the principle of opening another attacking front when you have the advantage across the board. Of course, Deep Blue didn't follow such general principles, at least not without breaking them down into smaller evaluation values first. Piece mobility is one thing computers can be programmed to understand quite well. I think the commentators were surprised because opening a second front in the position seemed like a strategic idea, and therefore unexpected from a machine. But humans are far more likely to fixate on a plan than a computer. Deep Blue looked at every single position afresh, with no dogmatic attachment to previous decisions or anything at all. This is one reason machines so often surprise us with their moves. Even Grandmasters often fall victim to rote thinking: "Having played A, now I must play B." A computer doesn't even know it had just played A; it only ever cares what the strongest move is at the moment. This could be a weakness sometimes, especially in the early phases, which is why they needed opening books. But, overall, their amnesiac's objectivity made them excellent analysis tools—and dangerous opponents.

An old chess joke that has been around since long before my time goes like this: A man is walking through the park when he sees a man playing chess against a dog. "That's amazing!" says the spectator.

"What's so amazing?" says the chess player, "I'm beating him three games to one!"

I was reminded of this joke when my colleagues and I began to analyze the Deep Blue match deeply for the first time to prepare this book. With the objectivity of time and using modern chess engines that are far stronger than Deep Blue, we made many interesting discoveries. It turns out that the commentary about this f-pawn move and several others like it set a pattern that was repeated in nearly every article and book about the match—a pattern that afflicted my thinking as well. It was the mistake of assuming that moves that were surprising for a computer to make were also objectively strong moves. We would often be so impressed that Deep Blue had made a certain move, a move of the type we hadn't seen computers make before, that it would influence our opinion of its actual quality.

For example, this move 26.f4 in game two is given a "great move" exclamation point annotation in several books on the match, but analysis shows that it's far from the best move in the position. Instead, it could have retreated its bishop and tripled its pieces on the a-file, gaining a dominant position without giving me any counterchances. Indeed, this superior plan is the first choice of any good engine today in just seconds. But like that chess-playing dog, it was so amazing to see Deep Blue playing as it did that the mediocre quality of its play was often overlooked. During the match, this had quite an effect on me. I became so concerned with what it might be capable of that I was oblivious to how my problems were more due to how badly I was playing than how well it was playing.

Still under the influence of my anti-computer delusions, I responded passively. Deep Blue continued to take over the board as I defended miserably. I passed up one last chance for an active defense on move thirty-two, still hoping it wouldn't find a way to break through decisively. It built up the pressure, playing with the patience of a Karpov. As I sat there suffering, Deep Blue surprised me with a very long think on move thirty-five. It usually played either very quickly or always within the range of three to four minutes per move. Here it sat for five, then ten, and then fourteen minutes before making its move. It was quite distracting and I thought maybe it had crashed. Apparently, it had gone into what the developers called its "panic mode" that kicked in when its evaluation dropped dramatically in a main variation.

On move thirty-six, it had the opportunity to invade my position with its queen, likely leading to the win of two pawns. I saw that this might give me the opportunity for a desperate counterblow in the center. Would greed once again be a computer's undoing?

Much to my dismay, Deep Blue again refused to play like a machine. Instead of grabbing the pawns, it played a bishop move that felt like the last nail in my coffin. After nearly four hours of torment in the sort of passive position I despised, it had somehow become even worse. A fatalistic depression set in and I could barely make the next few moves as white's queen and rook invaded on the a-file. My only hope was to establish some sort of blockade, but I couldn't see any way to achieve it. I gave a last check with my queen, almost out of spite, and barely even noticed that Deep Blue moved its king out of check by moving into the center instead of the more natural retreat toward the corner.

On move forty-five, it attacked my queen with its rook and it was all over. My queen couldn't escape without abandoning my bishop. I could sacrifice the bishop to get in a few desperation checks with my queen against white's open king, but that also looked hopeless. If computers are good at anything, it's seeing long sequences of checks, the most forcing move in the game. After such a powerful performance, it was inconceivable that Deep Blue would allow its king to be chased around for a draw when it had passed up simple ways to secure it.

The entire game had been a demoralizing experience and I just wanted to get as far from the board as possible. My mind was already racing, wondering how in the hell the time-wasting computer from game one had achieved this positional masterpiece in game two. That I was already thinking about anything other than the game was a typical human frailty that we just cannot avoid. It felt physically painful to keep looking at what I was sure was a totally lost position. I wanted to resign with at least a little dignity left and to save some energy for the next game instead of continuing in a hopeless cause.

I resigned and stormed away from the board, replacing my disgust with anger as quickly as I could. I was in no mood to face the audience or the commentators or anyone else. My mother and I left the building with no delay, leaving the Deep Blue team to their moment of glory.

I wasn't there to hear it, but the rave reviews rolled in for Deep Blue's performance in the auditorium, the transcripts show. Seirawan, frequently critical of the machine's play, said of the game, "I would be proud to have this one." Ashley called it "a gorgeous game" and he and Valvo praised the "anaconda" style of how it squeezed my position to death in a most uncomputerlike way. An audience member asked if it was the best game ever played by a computer and it was hard to disagree with the answer, that it was the best game a computer had ever played against Kasparov.

The comments from the elated Deep Blue team reflected their feeling that all their hard work over the previous fourteen months had finally paid off. Hsu: "This year it had a better understanding of chess and some of the subtleties of chess, and that showed up in this game." Benjamin: "The gratifying thing about it is that this is a game that any human Grandmaster would be proud to have played for white." This curtain call closed with a line that every newspaper writer loved when David Levy asked how Deep Blue had gone from making "a few dubious moves" in the first game to playing "like an absolute genius." C. J. Tan replied, "We let it have a couple of cocktails!" and brought down the house.

I HAVE NEVER BEEN much of a drinker myself, but I could have used something stronger than the hot tea I had that night as we looked over the game. It's always difficult to force yourself to review your losses, especially when you have to psych yourself up to go back out and fight hard in the next game. In a tournament, you don't play the same player twice with the same color, so an immediate postmortem isn't essential. In a match, you are going out against the same opponent day after day and it's important to see if anything can be gleaned from each game that might be useful in the next. This was especially true against Deep Blue, since these were the only games we had to go on.

Some losses are also much harder to take than others, and this was one of the worst I had ever experienced. It made me question everything: the dramatic increase in the quality of Deep Blue's play, the decision to play anti-computer chess instead of my own game, how I had been fooled into believing I would have some of Deep Blue's games to study before the match. Our analysis of the conclusion of the game didn't make me feel any better. How could a computer that had played in such an awkward, materialistic, and computerlike way in game one refuse to win material when it had the chance? Our engines never even considered Deep Blue's surprisingly patient moves at all.

After criticizing others for attempting to psychoanalyze me I won't make the same mistake, and I will stick to sharing my honest feelings at the time. I am aware of the mental defense mechanisms competitors use to deal with defeat, but I also know that they are a bit like particle physics: if you observe them too closely they don't work the same. I needed to recover my confidence for game three and for the rest of the match or it would be hopeless. I was confused and in agony and I was taking it out on everyone, especially on myself.

Little did I know that night that if it was already going to be very difficult to recover after such a loss, it was about to become impossible. My team—Yuri, Frederic, Michael, and Owen—and I were walking to lunch down Fifth Avenue the next day when Yuri approached me with the face of a man about to tell someone that a close family member has just passed away. "The final position of yesterday's game was a draw," he told me in Russian. "Perpetual check. Queen to e3. Draw."

I stopped dead still on the sidewalk with my hands on my head for a moment. I looked at each of them since they had all obviously known this and had been debating if, when, and how to break the news to me. They could barely meet my gaze, knowing how horrified I was by the news. I had lost one of the worst games in my life in front of the entire world and now I was finding out that I had resigned in a drawn position for the first time in my life. I was in disbelief, a feeling that I was becoming all too familiar with in this match. A draw?!

The Kübler-Ross model, better known as the five stages of grief, is a set of emotions that terminally ill patients and others experience after they receive terrible news: denial, anger, bargaining, depression, and acceptance. I spent the rest of lunch in a form of incredulous denial, staring at the walls for a few minutes running the variations through my head before I began hammering my poor team with questions. "How could Deep Blue miss something so simple? It played so well, it played Be4, it played like God, how could it miss a simple repetition draw?"

To psychoanalyze just this once, with twenty years to cycle through the stages, this was also me saying to myself, "My god, how could *I* miss something so simple?" When you are the world champion, the world number one, any defeat can be viewed as self-inflicted. This is not exactly fair to my opponents, many of whom would count their victories over me as the pinnacle of their careers, but after such an incredible revelation I wasn't in the mood to be fair to anyone.

The discovery had been made thanks to the power of the Internet to connect people around the world. Even before I resigned game two, the millions of chess players who were following the match got to work analyzing it and sharing their results. By morning, these armchair analysts, also armed with strong engines, had demonstrated that Deep Blue would not have been able to win the final position had I played the best moves instead of resigning. This unbelievable news was verified by my team that morning before they broke the news to me. The queen infiltration I discarded as desperate and pointless was, in fact, a saving resource. The white king would not be able to escape the checks by my queen, eventually resulting in what we call a three-time repetition of position draw. Several of Deep Blue's final moves had in fact been blunders that would have let its brilliant victory slip away had I only been alert to the opportunity.

It was a crushing blow, as if I had lost the game twice. Resigning in a drawn position, unthinkable! I would never have given up so pathetically against any human player in the same position, of that I was sure. I had been so impressed by Deep Blue's play, so demoralized by the way the game had gone, so annoyed at myself for letting it happen, and so sure that a machine would never commit such a simple mistake.

Against another Grandmaster, there is the assumption that he sees roughly what I see, and is unlikely to be sure of anything I am not sure of myself. But against a computer that could check 200 million positions per second, and had just played a powerful game against the world champion, the assumptions were different. I couldn't play normally; I had to give the machine the benefit of the doubt in certain positions. For example, if I thought I might have a powerful sacrifice leading to checkmate, I could be nearly positive that there was a flaw in my calculations because a strong computer would never allow such a thing. This was a necessary heuristic for a human playing against a machine: if it is allowing you to play a winning tactic it's probably not winning at all. This can help you save energy, but in this case it led to the worst blunder of my career.

The worst thing that can happen to you in a match is to let a loss cost you more than the one point. If your mental equilibrium is knocked off kilter, it can ruin your concentration and the losses can pile up quickly. A typical antidote is to try to gain a quick draw after a bad loss, as a way to steady the ship. But in a short match I could not afford to waste one of my remaining turns with the white pieces. And the Deep Blue team was unlikely to accept any draw offers in equal positions. After all, their player wasn't getting tired and would not be humiliated by the news that it had blundered and allowed a drawing variation in game two.

The furor about what had happened quickly spread through the media and dominated the headlines. I dreaded having to face questions about my early resignation. What could I say? Continued attention on game two would only serve to make it impossible to put it behind me, ruining my focus for the rest of the match. There were still four games to play, and I didn't feel like playing chess anymore. I didn't know my opponent at all. Was it the computer that made weak pawn moves in game one? Was it the strategic mastermind that had played game two like an anaconda? Or was it buggy and error prone, capable of missing a relatively simple repetition draw? This intense confusion left my mind to wander to darker places. IBM had made it clear they wanted to win at any cost. Would some sort of outside interference explain why Deep Blue had changed its character so radically?

Nor did I feel comfortable with myself. How could I have played such a terrible opening? Was I getting bad advice or simply making bad decisions? What should I change? How, how, how could I have resigned early?

WITH ALL THAT boiling away in my head, I had to sit down and play game three. The first move was another first for me, the sad push 1.d3, moving the pawn in front of the queen one square instead of the usual two. This was anti-computer chess taken to the extreme, getting Deep Blue out of its book and hoping to outplay it in early maneuvers. This strategy had paid off in the first game, although it was very much on my mind that I might no longer be facing "that" Deep Blue at all.

Looking at the game today, I'm a little surprised I played as well as I did considering what had transpired over the previous twenty-four hours. I didn't get much out of the opening after passing up a strong queenside expansion idea for reasons that I still cannot explain. Twenty years later, analyzing games three to six of the rematch feels more like analyzing the games of a stranger, not my own. I generally have excellent recall of what was going through my mind at different points in a game, even games I played decades ago. That's not the case with this match because I simply wasn't myself and my mind was not properly in the games.

Deep Blue's play in game three did not impress either, but it wasn't clear how I was going to make progress. I grabbed the opportunity to sacrifice a pawn to gain some pressure, burying black's bishop in the corner, although I was already becoming sure it wasn't going to be enough to tip the balance without an error by Deep Blue. Another moment that showed I was far from my best was when I forced the exchange of queens instead of playing my rook deep into black's position. It wasn't much of an improvement, analysis shows, but it would have been much more in keeping with my style. I was playing scared.

Deep Blue showed good alertness not to allow its pieces to become too passive, quashing my last hope to trick it into a passive defense where I could squeeze it at my leisure. My pieces achieved a superficial domination of the board, but without enough potential activity to convert it into a victory. Finally, Deep Blue gave back the pawn to reach a sterile endgame and a draw was agreed a few moves later. The game was over and an even tougher battle was about to begin.

I knew that the postgame press conference was going to feel like it was a year long as I was interrogated about Deep Blue's strong play in game two and my early resignation. I also knew that if I was going to play the second half of the match at a respectable level I could not allow myself to be put on the defensive. I have always believed that the best defense is a good offense, and that goes for chess, for politics, and for press conferences.

How could the tame game three compete with the fireworks of game two? A bad opening by the champion, brilliant positional play by the machine, a stunning coup, a shocking blunder, and a postgame revelation that shook the chess world. And looking through the transcripts, the commentators had been discussing little else all day. Frederic was there to regale them with the story of breaking the news to me, dramatized in his best style as usual, saying that they had decided to tell me, since otherwise the first taxi driver I met would do it anyway.

Seirawan, the only world-class Grandmaster on the commentary stage, was sympathetic to my plight and attempted to convey to the audience what I had gone through, and was still going through, even as I labored at the board during game three. "He convinced himself he had a lost position so he resigned. We humans get depressed. . . . Chess professionals are very proud persons. They are artists, they take their art very, very seriously. To play a great game is very meaningful for a player's career. To resign a drawn game is unthinkable. I mean, I would just torture myself mentally. How does Garry recover from something like this?"

Based on its apparent lack of comprehension of the position, it's likely Deep Blue would have failed to win had I played on and found the best moves myself. Hsu writes that he later analyzed that final position in shock, realizing it should have been drawn. But the story took another twist years later, as analysis went deeper than we had time for then. Today, strong engines show that white was still close to winning.

If you'd like to check it at home, simply put in the final position of the game and see what a chess program thinks. Even if it's a free engine on your phone, it will likely show a plus score for white of nearly a full pawn in value. While you are there, look at the position before Deep Blue's final move, 45.Ra6. Today's machines see instantly that simply exchanging queens is crushing for white. Amazingly, the final two moves of the mighty machine's masterpiece were serious mistakes. But I made the last and worst blunder by resigning.

Prior to game three, I had requested the machine's logs, the printouts, for the moves in game two that I felt defied explanation, including the final move that turned out to be a blunder that could have allowed a draw. My request was declined by Tan, who said that we could use the logs to figure out Deep Blue's strategy, although I failed to understand how the last move blunder could be revealing in that way. We petitioned the Appeals Committee for the printouts, hoping that at least there would be a permanent record for public examination. After some negotiations, Tan said he would give them to Ken Thompson, who was on the Appeals Board and was operating as a neutral tech supervisor, as he had in Philadelphia. That also did not happen despite repeated requests, and the saga over the printouts was under way.

On my way to the dreaded press conference I decided that I was going to say what I thought, consequences be damned. I had earned my right to an opinion and if I was uncomfortable and confused by what I had experienced, I would say so. In order to play chess, I needed to get through the denial and confusion and move into a cleansing anger. Years of all sides trying to spin what happened have resulted in a revisionist history that portrays me as a sore loser trying to explain away my loss with wild conspiracy theories. To the sore loser accusation, I have already confessed my guilt. As for the wild theories, the transcripts of the press conferences show the truth. It was far more a matter of my expressing my doubts and frustrations. I didn't know what had happened and I admitted it. I couldn't understand how a machine could play so well and then make a blunder that seemed elementary, and I said so. I challenged them to explain it to me and to the world, to release the printouts and remove all the doubts, but they wouldn't. Why not?

After I expressed my bafflement several times, Maurice Ashley asked me specifically if I was implying that there had been "human intervention" in game two. "It reminds me of the famous goal Maradona scored against England in 1986," I said. "He said it was the hand of God!"

The audience laughed, as I intended, although I didn't realize that the largely soccer-ignorant American audience wouldn't grasp the reference. Argentine soccer legend Diego Maradona had scored a goal against England in the quarterfinals of the 1986 World Cup in Mexico. It wasn't clear at the time to anyone but the players on the field nearby, and certainly not to the arbiter, that Maradona had actually punched the ball into the net with his left fist, as replays would show. When asked about it after the game, won by Argentina 2–1, Maradona responded with the brilliant evasion, "It was a little with my head and a little with the hand of God." Until I could be shown evidence otherwise, unseen forces were one possible explanation for things that I could not explain.

Benjamin and I sparred a little at the press conference about what Deep Blue could and could not see during the key moments I wanted the printouts from. C. J. Tan tried to calm the waters by replying affirmatively to a suggestion by Valvo that we "get together in a lab after the match" to go over the position from the end of game two. "Sure, after the match we'll be glad to have Garry come up to our lab and continue our scientific experiment with him."

I had started calming down, but this attempt at an olive branch only got my adrenaline surging again. Hadn't Tan himself told the *New York Times* that "the science experiment is over"? If this was about science, why not release the printouts to clear away the doubts? I replied that if we were talking about "the purity of the experiment, then one would like to have both opponents under equal conditions." Campbell also implied to the press that the curtain would be pulled back as soon as the match was over, saying, "He doesn't know how we did what we did, and at the end of the match, we'll tell him."

CHAPTER 10

THE HOLY GRAIL

Since I am now in my mid-fifties and must take care with my blood pressure, allow me to leave that bitter scene behind for a moment before returning for the final games of the rematch and a closing press conference that made the one after game three look like a child's tea party in comparison.

There is a long and ugly history of recriminations and accusations of foul play and worse during world championship matches. The most popular anecdotes are trotted out in every layman's book about chess because they are amusing from a distance. Fischer's protests about the cameras in the playing hall in his 1972 match against Spassky led to him forfeiting the second game and to the third being played in a little room instead of the main stage. Karpov and Korchnoi often feuded viciously during their matches, especially their 1978 world championship in the Philippines. Karpov had a psychologist, some say a parapsychologist, named Dr. Zukhar on his team who stared at Korchnoi during the games. A series of dueling protests moved the man around the room in nearly every game. Korchnoi retaliated by inviting some American members of an Indian sect, who meditated and stared at the players and Karpov's man. There were protests and investigations about the chairs—including having Korchnoi's dismantled and x-rayed—Korchnoi's mirrored glasses, and Karpov's yogurt.

The 2006 world championship match between Vladimir Kramnik and Veselin Topalov sank to new lows—all the way to the plumbing fixtures. After accusations by Topalov's camp that Kramnik was spending a suspicious amount of time in his personal bathroom during games, the organization closed it, leading to Kramnik forfeiting the fifth game in protest in a scandal quickly dubbed "Toiletgate" by the chess media. (Kramnik went on to win the match anyway.)

My epic rivalry with Karpov was not immune to such adventures, naturally. In the 1986 rematch, Karpov repeatedly displayed an almost magical intuition regarding my opening preparation. He met several of my novelties almost instantly, with the strongest responses, and seemed completely prepared even for lines I could not have been expected to play. I felt that the only way this could keep happening was if someone on my own team was sharing my preparation with Karpov. Two of my team members ended up leaving, though not before I lost three games in a row. A later article by one of Karpov's own team includes remarks about how Karpov had spent a sleepless night analyzing a variation "he was sure would occur" in our next game, despite it being a completely different opening from my previous two games with white, which I won. Needless to say, his premonition was correct.

In sum, you can either believe that there is a great deal of treachery at the top level of chess, that some Grandmasters are as paranoid as the stories say, or that gamesmanship and off-the-board maneuvers are a standard element of an all-out psychological war. Or you may select "all of the above" and join the consensus.

My next clarification is about the dangerous words Ashley floated into the air at the press conference: "human intervention." I have spent twenty years dealing with the many loaded meanings of this phrase, although I did not coin it and my insinuations were more complex. A certain amount of human intervention was allowed on Deep Blue's behalf during the match. They were allowed to fix bugs, reboot after crashes, and to change its book and evaluation function between games, for example, and they did so. Later human-machine matches would limit this sort of activity, judging it as an unfair advantage for the computer.

There were at least two crashes of the machine during play, requiring a manual restart. According to the Deep Blue team, this happened in game three and game four. While neither incident seemed relevant to them because it didn't affect Deep Blue's next move, having to ask Hsu what was going on during a tense endgame in game four was far from ideal. And, as was subsequently pointed out to me by several chess programmers, a system restart changes everything from the perspective of reproducibility. The memory tables the machine uses to retain positions are lost and there will never be a way to confirm that the machine would repeat the same moves.

Putting allowed user actions aside, most people take the idea of human intervention to mean that Karpov or some other strong Grandmaster would be hiding in a box somewhere making moves in the style of Wolfgang von Kempelen's chess-playing hoax automaton, the Turk. But in the modern day, with all the backups and remote access points, it wouldn't require a chess master dwarf hiding inside the big black box. An amusing thought, but not really the point. Simply rebooting the machine, or triggering an event that forced Deep Blue to take extra time in a tricky position could be enough to make a big difference. Remember that the Deep Blue prototype in the 1995 Hong Kong tournament had to be restarted during its key game against Fritz, and it came back online with an inferior move. Bad luck, but it could also have come back with a superior move, especially if, for example, it had been programmed to take extra time after a crash.

On September 15, 2016, I was in Oxford to speak at the Social Robotics and AI conference, and I jumped at the chance to meet Noel Sharkey. From the University of Sheffield, Sharkey is one of the world's great experts on AI and machine learning, and he's currently involved in various projects on ethical guidelines and the societal impact of robots. But he's best known in the United Kingdom as an entertaining expert and head judge on the popular television show *Robot Wars*. We only had a short time to speak during a lunch break before his conference keynote. I wanted to talk about machine learning and his United Nations debate on robot ethics. But he wanted to talk about Deep Blue!

"I've been annoyed about it for years," he told me. "I was very excited about the prospect of an AI system beating you but I wanted it to be a fair contest and it wasn't. The crashes? All the connected systems they put in? How do you monitor that? They could change software or hardware between moves. I can't say IBM cheated but I can't say that they didn't. They certainly had the opportunity. Forget it! If I had been adjudicating. I'd have ripped out all their wires, put a Faraday cage around Deep Blue, and said, 'Okay, now play, you're on your own.' Otherwise I'd have forfeited the damn thing in a second!" The mental image of Noel Sharkey ripping network cables out of Deep Blue made me sure I'd want him on my team no matter who I was playing against.

Lastly, the argument that IBM would never do or allow anything inappropriate in order to help Deep Blue's winning chances was popular at the time but sounds almost quaint today. Nineteen ninety-seven was still four years before the Enron scandal rocked the corporate world, exposing the American energy giant as an empire of malice and fraud. It was a Watergate moment for the corporate world, and a preview of the revelations that came out of the 2007–08 financial crisis. I'm not putting a chess match on the scale with a catastrophic financial meltdown, of course. I make the point because after Enron, people stopped telling me that "a big American corporation like IBM would never do anything unethical." Especially after they found out how much IBM's stock price went up after the match.

Thanks to the honesty of Miguel Illescas, we do know that IBM was willing to push the boundaries of ethical behavior to improve Deep Blue's chances in any way. In his 2009 *New In Chess* interview, he shared this remarkable revelation: "Every morning we had meetings with all the team, the engineers, communication people, everybody. A professional approach such as I never saw in my life. All details were taken into account. I will tell you something which was very secret. Well, it's more of an anecdote, because it's not that important. One day I said, Kasparov speaks to Dokhoian after the games. I would like to know what they say. Can we change the security guard, and replace him by someone that speaks Russian? The next day they changed the guy, so I knew what they spoke about after the game."

As he says, perhaps not that important in practice, but it's a bombshell in exposing the lengths IBM went to in order to gain any competitive advantage. I can hardly imagine the scandal that would have erupted had it been revealed during the match that IBM had hired Russian-speaking security personnel stationed in my personal rest area specifically to spy on me and my second during the match, but it would have been ugly.

After saying all of that, we come to my own confession. On what mattered most, on what really destroyed my composure, I was wrong and owe the Deep Blue team an apology. The moves in game two that left me with a lost position and crushed morale were unique only for the time. Within five years, commercial engines running on standard Intel servers could reproduce all of Deep Blue's best moves, even improving on some of the "humanlike" moves that so impressed me and everyone else at the time. The engine on my laptop today slightly favors the "shockingly humanlike" move 37.Be4 from game two in less than ten seconds, although it rates it nearly equal to the queen sortie I had expected because 37.Be4 turns out not to have been as superior as we all believed at the time. Had I played better defense instead of collapsing and resigning, game two would have been considered a very impressive game for a machine but nothing more, no matter the eventual result.

This also highlights why it was so critical that I never saw a single game of Deep Blue's before the match. Had I seen it make a single move demonstrating the uncomputerlike positional approach of game two's Be4, for example, or the surprising h5 pawn push from game five, my play and my reactions would have been completely different. Keeping Deep Blue completely hidden was the strongest move of the match, but it was made by IBM, not by either of the players.

In turn, by understanding now that Deep Blue was very strong but still making plenty of inaccuracies, the fact that it missed the perpetual check draw at the end of game two becomes more comprehensible as well. Still strange, considering its powers of calculation, but no longer inconceivable. If I had had any way of knowing this during the match, perhaps the story would have turned out differently, but I'm not sure. My premature resignation in game two and the intense shame and frustration it produced in me were what made it nearly impossible to play.

I have my regrets, but I was not wrong to be shocked and confused at the time. In 1997, Deep Blue's play was completely inexplicable to me, and IBM went to great lengths to keep it that way. Maybe they really didn't have anything to hide, but they realized it couldn't hurt if they acted as if they did, stoking my suspicions. They continued to stonewall us on releasing the printouts from game two, which, if they had showed nothing amiss in the eyes of Ken Thompson, could have relieved some of my stress about what was going on behind the scenes.

My agent Owen Williams told the organizers before game four that if Thompson did not receive the game two printouts, he wouldn't be able to appear as a member of the Appeals Board. IBM took this as a warning that I might not appear either if Thompson didn't, and they warned the media that there might not be a game that day. Thirty minutes before the game started, we got a message from Newborn saying that the printouts had been given to the Appeals Board, but when we arrived at the thirty-fifth floor, Thompson said that they were only for one move, 37.Qb6. Without the other moves to show context, this was useless.

This secretive and antagonistic behavior manifested in other ways as well, as reported in the *New York Times* after game five: "One reporter, Jeff Kisseloff, who had been hired by IBM to report on the Kasparov team for the match Web site, lost his reporting privileges after he included damning comments about Deep Blue from the champion's supporters in his report. IBM also engaged Grandmasters John Fedorovich (*sic*) [a.k.a. Fedorowicz] and Nick DeFirmian to work on openings with Deep Blue, though no one on the Deep Blue side has said so publicly, even when asked directly in a news conference about additional help. It was Mr. DeFirmian who confirmed his involvement and that of Mr. Fedorovich, but declined to discuss it, he said, because IBM had insisted he sign a secrecy agreement."

All of this prompted my mother to say, "It reminds me of the 1984 world championship match against Karpov. You had to fight Karpov and also the Soviet bureaucracy. Here we are, thirteen years later, and you have to fight a supercomputer and also a capitalist system using psychological warfare." (If her use of "capitalist" sounds like a Marxist anachronism, remember what the first match did for IBM's stock price!)

THE GAMES had to continue at the board as well, and I had black in game four. Obviously I would not repeat the disaster of game two, in which the typical roles of human-machine chess had been reversed. The computer had built up a dominating position with strong strategic play while I had been reduced to defensive shuffling. But when the machine finally broke through to convert its advantage, it made a tactical slip that could have been immediately exploited to lead to a shocking forced draw. (As everyone still thought at the time.) It was the same pattern as countless games since the first human-machine contests, only the computer and human had switched positions. In games four and five, the players would resume their normal roles.

I returned to a flexible defensive system in the fourth game and achieved a solid position after several diffident moves by Deep Blue. It still occasionally showed the downside of not being able to logically connect moves the way a human does. It advanced its pawns on the kingside and then seemed to forget about them as it found other alternatives, giving a strange impression. Again, there are advantages to this extreme objectivity, but there is a reason we say that a bad plan is better than no plan, at least in human chess. If you have a plan and it fails, you learn something. If you act aimlessly, from move to move, from decision to decision, whether in politics or business or chess, you don't learn and will never become anything more than a skillful improviser.

The machine was pushing hard, too hard, and by so doing it created weaknesses in its own camp. On move twenty, I played a strong pawn sacrifice to free my pieces and to turn the tables. The machine again made a couple of strange moves that the commentators, at least the human ones, were quick to dub "ugly" and "pointless." GM Robert Byrne wondered, "How can it be very strong one day and loony the next?" And perhaps they were loony to the commentators, but I had already come to appreciate that Deep Blue had a knack for making its moves work, however ugly they might be to a Grandmaster. This makes sense, because even if a machine doesn't employ the goal-oriented strategy that humans use, if it evaluates a move as the best it's because something in its evaluation likes it and the positions resulting from it. It's an alien version of how Grandmasters have different styles. A move made by former world champion Tigran Petrosian, famed for his defensive skills, might look completely pointless to an attacker like me. Indeed, that move would in effect be weak if I made it, but it was strong for Petrosian because he understood it and what would come of it. Deep Blue could still make genuinely weak moves and pointless moves, of course, but it was strong enough that its computerlike inconsistency often worked out fine for it.

In another terrible disappointment, game four turned into another example of this. I missed one good attacking chance, but still had the clear upper hand all the way through to the endgame, only to find that the machine had a series of incredible drawing maneuvers that I could never have foreseen. Even today, looking at the game after move thirty-six, I cannot believe I failed to win that position, and even more incredibly, that the position might not even be objectively winning at all. With two rooks and a knight each, and a scattering of pawns, every aspect of the position was in my favor. My pieces were more active, its pawns were isolated and vulnerable. Even my king was better placed for the endgame. I estimate that I would win that position against a very strong Grandmaster four out of five times.

It was almost as if Deep Blue was taunting me by getting as close to losing as possible before coming back to draw. Material on the board slowly dwindled and I was finding it hard to calculate clearly as I began to tire. The forced win I was so sure was just around the corner never stopped being just around the corner. Commentators and later analysts were as surprised as I was, and kept looking for mistakes in my play that had let Deep Blue off the hook in the ending. But while perhaps I did not play flawlessly, it appears that there was simply no win to be had. Any strong player could explain why black's position was clearly superior, but even Grandmasters with strong engines have failed to demonstrate how to win it. It was another demoralizing and exhausting day at the board.

After the game, I asked Frederic if he thought Deep Blue had used its secret weapon to help it achieve the miraculous draw. There were rumors that the machine could access endgame tablebases during its analysis, and, if so, I would have Ken Thompson to blame for it. In 1977, Thompson showed up at the World Computer Chess Championship with a new creation, a database that played the king and queen versus king and rook endgame perfectly. (KQKR is the abbreviation.) It wasn't an engine; there was no thinking required. Thompson had generated a database that essentially solved chess backwards, what we call retrograde analysis. It started from checkmate and worked its way back until it contained every single possible position with that material balance. Then it worked out the optimal move from every one of those positions. For example, in KQKR, for the side with the queen it always played the moves that led to checkmate quickest. For the side with the rook it always played the move that delayed checkmate the longest. It didn't play like a god, it *was* God. Or more accurately, the goddess of chess, Caissa!

It was a revolutionary contribution to computer chess, where the subtleties of endgame play had long been a machine weakness. A human can look at a pawn endgame and see instantly that if one side has two pawns versus one on the same side, he can create a passed pawn that will become a queen. That might take fifteen or twenty moves to actually happen on the board, but you don't have to calculate them all to know what will eventually occur. A computer, on the other hand, does have to calculate all the way to the pawn queening to see the truth in the position, and that was often far too deep even for strong engines to reach.

With tablebases, all that started to change. Instead of calculating all the way, a machine only had to reach a tablebase position in its calculations to know if it was winning, losing, or a draw. It was like gaining second sight. Not every chess game reaches an endgame, so their utility was limited, but as tablebases grew bigger and bigger, incorporating more and more pieces and pawns, they became a powerful new weapon in the computer arsenal.

Thompson's endgame databases were also the first computer chess innovation to have an impact on human chess. When he started with KQKR, he challenged Grandmasters to play against it, to see if they could win with the queen against his database. Keep in mind that it was generally considered not that difficult for a strong player to win queen versus rook; the general algorithm was taught in every endgame book. Incredibly, the machine showed how hard it really was, and it did it by playing moves that were inexplicable even to Grandmasters.

Six-time US champion Walter Browne lost a bet with Thompson when he failed to beat the database in under fifty moves—the amount of moves the rules of chess allow you to try to win such positions before the defender can claim a draw. Shocked, Browne, ever the gambling man, studied for a few weeks and returned for another try, mating it in exactly fifty moves and getting his money back. The position was actually a win in just thirty-one moves with perfect play, according to the database. For the first time, humans were being exposed by computers as far from perfect chess players.

The massive data storage required for each new piece added at first made tablebases impractical for most engines. One common set requires 30 megabytes for all positions with four pieces, 7.1 gigabytes for all positions with five pieces, and 1.2 terabytes for all positions with six pieces. Their use became commonplace as new data generation and compression techniques came along, and as hard drives kept getting bigger and bigger.

Just as the search tree from the beginning of a chess game grows too quickly to ever solve chess from the start, tablebases are far too huge and too difficult to generate to ever solve chess from the end. Theoretically, a thirty-two-piece tablebase could be generated, but we cannot even conceive of how much storage space it would require. Seven-piece bases only started to appear in 2005, due to the massive computing resources they require to generate and store. There are now full sets of seven-piece tablebases that take months to generate and occupy 140 terabytes. Now accessible online, they were originally generated by Russian researchers Zakharov and Makhnichev using a Lomonosov supercomputer at Moscow State University.

These have revealed some fascinating things about the complexity of chess while also refuting centuries of chess analysis and studies. For example, the longest mating position for seven pieces is KQNKRNB (king, queen, and knight versus king, rook, knight, and bishop). If configured just so, it takes exactly 545 perfect moves on both sides to force checkmate. More practical and well-known positions have also had to be reevaluated. It was assumed for a century that in some positions it was impossible to win with two bishops against an ideally placed lone knight, but the tablebases showed this was false.

There is a long history of chess studies and problems, in which the composer artfully arranges the pieces and presents the reader with a stipulation, usually "white to play and win" or "white to move and mate in three." These are often found in the chess columns of local newspapers—if newspapers still have chess columns (and if we still have newspapers). Many of them look simply impossible and their solutions often reveal great wit and beauty. Databases care not for such things and many compositions have been refuted by the machines.

In a few cases, how the databases play some common positions can be useful for the human player to study, but this is rare. We need useful patterns and heuristics like "put your rook behind the passed pawn" or "keep your rook near your king when defending against the queen" in order to play. Tablebases generally provide no help in how to make these endgames easier for humans to understand. Even to me, 99 percent of tablebase moves in some positions are completely incomprehensible. I have flipped through several six- and seven-piece endings that require over two hundred moves to solve, and often the first one hundred and fifty moves looked like nothing was happening at all, revealing no pattern I could grasp. Only as mate came within forty or fifty moves could I start to see method in the machine madness.

It was one thing to face a giant opening database that had been prepared by a team of Grandmasters. It was quite another thing to play against an endgame database that played literally perfectly. Later human versus machine matches took steps to balance this part of the playing field as tablebases became bigger and more common. For example, in my 2003 match with Deep Junior, this line was added to the rules: "Should a position be reached which is in the machine's endgame databases and if the result from that position with correct play is a draw, then the game ends immediately." Otherwise, a game could become more of a strange form of solitaire than a competition.

Tablebases are the clearest case of human chess versus alien chess, and of the huge difference in how humans and machines achieve results. A decade of trying to teach computers how to play endgames was rendered obsolete in an instant thanks to a new tool. This is a pattern we see over and over again in everything related to intelligent machines. It's wonderful if we can teach machines to think like we do, but why settle for thinking like a human if you can be a god?

This question was on my mind when I looked over Deep Blue's incredible defense in game four. The rook endgame had been drawn with eight pieces on the board, too many for tablebases then or now to render a perfect verdict. But what if Deep Blue was accessing tablebases during its search? Could it be looking ahead and checking to see which positions were winning and losing in order to improve its evaluations? This "probing" of tablebases in the search later became standard for engines, but we weren't sure if Deep Blue was doing it. If so, it was cause for concern. Would I have to add some endgames to the realm of positions I had to avoid against Deep Blue?

According to papers published later by the Deep Blue team, the machine did have access to tablebases during the match, and indeed used them briefly in its search in game four, the only game that reached a simplified endgame position. Six-piece tablebases were quite rare at the time, so I was surprised to read that Deep Blue's contained "selected positions with six pieces" they had specially requisitioned from an expert.

Game four also included another crash, after I made my forty-third move. Every computer user knows what a crash is: your machine freezes, or the screen turns blue, and it's time to curse and reboot. I've had many laptops and projectors crash on me during my lectures, which gives me the chance to quip that it's because computers still hate me! But in discussing these events with experts, including one of the creators of the multiple computer world champion program Deep Junior, Shay Bushinsky, I realized how oversimplistic my understanding was. He pointed out that just about anything can take place during the recovery process, especially if it was a "controlled crash" instead of a catastrophic halt. Programmers often insert code that will restart all or part of their program's processes under certain conditions. In fact, this is what happened to Deep Blue, according to Hsu's book *Behind Deep Blue*. He calls them "self-terminations," not crashes, and describes it as a "piece of code that monitored the efficiency of the parallel search and terminated the program itself if the efficiency dropped too low."

This is a remarkable admission because it says that these distracting crashes—sorry, these distracting "self-terminations"—were a feature, not a bug. Not exactly intentional, as in occurring on demand, but a working part of the system used to "clear the pipes" if Deep Blue's parallel processing system became clogged. This isn't to say that they directly improved Deep Blue's play, or that it would necessarily be unfair if they did, depending on the rules in place. But aside from annoying me during play as they fiddled with the machine, it made the games impossible to reproduce.

This was the biggest problem, according to Shay. "Once it crashes, the entire thing is kind of a sham because you can never confirm what happened is authentic," he told me over dinner near his home on a sweltering evening in Tel Aviv in May 2016. I was in Israel to give two lectures, one on education and another on the human-machine relationship, and took advantage to gain the input of an old friend and colleague who also happened to be a world-class expert on machine chess. "The move timing changes, the hash tables change, who knows what else? There is no way to say afterward, 'This is exactly why the machine made this move' with any conviction. That's not so bad in testing or in a friendly game, but in a high-profile competition, with millions of dollars at stake, it's unacceptable."

The game four crash took place on Deep Blue's move where, in a stroke of luck, there was only one legal move in the position. I had just checked its king with my rook and its reply was forced, so there was little concern this time that it would be aided or harmed by the reboot.

IBM CEO Lou Gerstner made a visit to the match during the game, though I doubt he was informed that his computer star had crashed again. All the great PR Deep Blue was providing IBM would have taken quite a blow had the media started asking about crashes or self-terminations. Gerstner gave his team a pep talk and told the press that the event was "a chess match between the world's greatest chess player and Garry Kasparov." Considering that the match was tied and Deep Blue's only win was in a drawn position that I had resigned, this seemed more insulting than accurate.

I felt completely drained, but we had two days off to prepare for the final two games of the match. I very much wanted to use my turn with white pieces in game five to make Gerstner eat his words.

We had already scheduled a special dinner for my team and friends for that night, although I really just wanted to go to sleep for ten hours. On the first rest day, we prepared a little for my black in game six. Then, on Friday, we started on game five and decided to stick with the anti-computer strategy that had done reasonably well in games one and three. The Réti Opening it would be. Meanwhile, we had asked that the printouts from games five and six be sealed immediately after the game and given to the Appeals Committee for safekeeping.

The opening of game five again showed the ups and downs of my anti-computer, anti-Kasparov strategy. I got the maneuvering position I wanted despite losing some time in the opening. I hadn't gained any real advantage with the white pieces, but there was still a long game ahead. Deep Blue's eleventh move was a surprising one, pushing its h-pawn forward two squares. The commentators thought this might be another case of Deep Blue making silly computerlike moves, but I wasn't so sure. It created a threat on the kingside and appeared to me to be less the move of a machine than one in the style of a very aggressive human player. It was early in the game, so there were many logical moves for black. Its choice of this surprising thrust at the edge of the board again had me shaking my head at what Deep Blue was capable of. I think I even glanced up at Campbell for a second after he played ..h5, as if to confirm it wasn't a slip by the operator.

It turns out that ..h5 wasn't very good and that I could have gotten a large advantage by moving my knight to the e4 square, but I responded weakly. Once again, it was a case of a strange but weak move by Deep Blue turning out to be more effective than a good move because of how it affected me psychologically. I never got a sense of what to expect, never felt sure of how I should play, and I let it ruin my concentration. And when these odd moves were combined with all the off-board conflicts, I also let my imagination get the better of me.

The position opened up while I searched for a way to secure an advantage. Analyzing today, I am again struck by how many opportunities I missed. I was at my peak as a player and as of this writing I have been retired from professional chess for over a decade. And yet some of my moves seem obviously bad to me, and analysis backs this up. As poorly as I played, I was lucky the match did not turn out even worse for me.

After some exchanges, it looked like the position was about equal. I didn't see how either side could play for a win. Then, to my delight, Deep Blue played a terrible queen move, allowing me to exchange the queens. Without the powerful queens on the board to generate threats, black's structural weaknesses became more prominent. Now I had targets I could go after the way I did in game two of the Philadelphia match.

It worked for a while; I was making progress as more pieces were exchanged. Just as in game four, I would look at this endgame and be absolutely confident I could win it against any human player. But once again Deep Blue defended aggressively, finding remarkable tactical resources to hold on. It brought up its own pawn and king to create threats against my king and I was forced to accept a pretty repetition draw with my pawn one square away from becoming a queen. I had seen the forced draw coming much earlier than the commentators, who were still under the impression I was winning until almost the last minute. For the second game in a row I felt shattered, certain I had squandered a winning opportunity and disgusted with the low quality of my play.

Before I left the board, I demanded that the printouts be turned over to the arbiter or Appeals Committee immediately. The room filled with people, much to the confusion of the spectators watching on the video screens. After more promises were made by C. J. Tan, who had earlier told the Appeals Board there wouldn't be any printouts until after the match, we went downstairs to discuss the game with the audience. Afterward, we went up for the printouts and no one was there. I went back to the hotel while Michael and my mother waited and tried to reach someone. The printouts would, finally, be delivered by arbiter

Carol Jarecki. (Deep Blue's full analysis logs wouldn't appear in public until several years after the match was over, when they were quietly uploaded to the website IBM created for the event.)

In the auditorium, I was again met with cheers. I was simply unable to feel buoyed by the crowd's support by that point, as nice as it was to hear. I felt like I couldn't see anything anymore. Even after it gave me several chances, I couldn't find the win I was sure was there, leading to another incredible escape by the machine. It was incredibly frustrating. That was an accurate assessment, analysis with modern engines shows. I had missed two good winning attempts and Deep Blue had again blundered badly, but yet again I had failed to exploit its mistakes. It turned out much later that I did miss a win in the game five endgame, not that this made me feel any better.

At the press conference, I was again frank about how impressed and surprised I was by some of Deep Blue's moves, especially the one that had elicited laughter from the commentators. I said, "I was very much amazed by ..h5. Many discoveries in this match, and one of them is that sometimes the computer plays very human moves. ..h5 is a good move and I have to praise the machine for understanding very, very deep positional factors. I think it's an outstanding scientific achievement."

I want to enter that statement in my defense for when I'm told I did not give enough credit to Deep Blue and its creators, especially since it turns out that ..h5 wasn't even a very good move! When the match ended the next day, and because of how it ended, I was in no mood to be flattering.

When asked about remarks by Illescas that I was afraid of Deep Blue, I was again candid. "I'm not afraid to admit I am afraid! And I'm not afraid to say why I'm afraid. It definitely goes beyond any known program in the world." At the end, Ashley asked me if I was going to try to win the final game with the black pieces and I replied, "I'll try to make the best moves."

IN A MATCH of many firsts and many records, game six of the rematch would add several others, none of them good for me. It was the shortest loss of my career. It was the first classical match loss of my career. It was the first time a machine had defeated the world champion in a serious match. As with exhibition games, such things acquire an asterisk in the record books when the game or match is against a computer, but I wasn't concerned about asterisks or my place in history. I had lost, and I hated losing.

The stories around the sixth and final game have grown in a way that I suppose is fitting for such a historic moment. It has acquired its own mythology, with different factions arguing for their interpretations. Rumors about what really happened in game six are passed around among the faithful like the shreds of a prophet's shroud.

The chess always came first with me, and so I very much wish the game itself were worthy of the moment and of so much attention. Losing a battle, even losing a masterpiece, would have been far more to my liking. Instead, it is little more than an ugly joke of a chess game, promoted into a historical artifact by circumstance.

The match was tied, 2.5–2.5. Should I play it safe and aim for a draw or should I risk everything and play for a win with black? With no rest day, I knew I would have no energy for another long fight of the sort that resulted from my anti-computer lines. My play was already shaky. I knew my nervous system very well from two decades of competition, and it would not withstand the strain of another four or five hours of tension against the machine. But I had to try something, didn't I?

It was the second time in the match that I played a "real" opening. The first time was the failed Ruy Lopez experiment in game two. This time I played the Caro-Kann, a solid positional choice that was a favorite of my nemesis Karpov, who used it against me several times in our games. I played it extensively in my youth, but decided fairly early on that the sharp Sicilian was far more in keeping with my attacking style. Deep Blue continued with a main line that I knew very well from having played it with white on numerous occasions. Perhaps Deep Blue's opening book coaches had a sense of irony, or maybe they just thought that if it was good for me, it would be good for their machine.

On the seventh move, still following the main line, I reached out and played my h-pawn one square, instead of the bishop move that usually precedes the pawn move. There were shouts of disbelief in the commentary room as Deep Blue responded instantly, crashing its knight into my position with a devastating sacrifice. My king was exposed, my pieces were undeveloped, and white had overwhelming threats. You can see on my face that I knew the game was already over. I went through the motions of trying to defend a position that would be very difficult against any Grandmaster and, I knew, was absolutely hopeless against Deep Blue.

I played another dozen moves on autopilot, barely registering what was happening. I ignored it when operator Hoane picked up the wrong bishop on move ten. On move eighteen I had to give up my queen and on the next move, facing further losses, I resigned. The whole game had taken less than an hour. The match was over.

If you can, for a moment, imagine what that moment felt like for me, take one extra step in my shoes and imagine then having to face hundreds of reporters and a large audience asking you about it. The press conference felt like a strange continuation of the game. I was in shock, exhausted, and bitter about everything that had happened on and off the board. When it was my turn to speak, I told the audience that I certainly did not merit their applause after what had happened in the final game, and I admitted that I had felt like the match was already over after I failed to win the endgame in game five. I said I was ashamed. I admitted that it had been a big mistake not to prepare for the match properly and to play my normal preparation, that my anti-computer strategy had not worked.

I reiterated both my praise and my concerns over Deep Blue's inexplicable moves, and threw down a challenge to IBM to let Deep Blue participate in regular tournaments, when, I promised, "I will rip it to shreds." I said I would play Deep Blue under any conditions, with the only caveat being that IBM could only participate as a player, not as sponsor or organizer. I announced I would play it again with my world championship title on the line.

When I read over the transcripts of the press conference to refresh my memory, I didn't think I sounded quite like the villain I was portrayed as afterward. I went on too long from sheer adrenaline, and repeated myself more than once. And I was far from gracious toward the victorious Deep Blue team in their moment of glory, and for that I must apologize.

But when I listened to the audio of the press conference, I could understand why they later said that I had "taken the joy out of it." Drained and disappointed, the anger and confusion are palpable in my voice. I cannot say that I regret speaking my mind, because it is my nature to say what is in my heart. But I could have waited until the next day, after some rest and contemplation. It is fair to say that I had failed to rise to the occasion in game six and then I failed again at the press conference.

So, what did happen in game six? When asked several times at the press conference I deflected the question: "It doesn't even count as a game." "I was not in the mood of playing at all, I have to tell you." "When you allow this piece sacrifice you can resign and there are many games played in competitive chess in which this line has happened, but I can hardly explain what I did today because I was not in a fighting mood."

This was all true, but it does not explain why I played the horrible 7..h6 instead of the normal 7..Bd6. Several competing theories have formed the mythology of game six. One, that I was so discombobulated and tired that I had transposed these routine moves by accident, playing them out of order. My defenders and friends have advanced this theory, which made its way into various news reports and books. Two, that I was trying to lure Deep Blue into a trap, based on some recent analysis in a computer chess journal that showed black could defend after the knight sacrifice. Three, that I played the Caro-Kann as a last-minute inspiration and didn't prepare, leaving me ignorant of this devastating blow.

Honestly, I find the suggestion that I blundered in my preparation to be more insulting than the idea that I suffered a complete nervous breakdown. Of course I was aware of Nxe6. I was also aware that it would be a killing move if Deep Blue played it against me in game six. I simply knew that it wouldn't.

Machines are not speculative attackers. They need to see the return on their investment in their search before they invest material. I knew that Deep Blue would decide to retreat its knight instead of playing the sacrifice, after which my position would be fine. I knew I didn't have the energy for a complex fight and that I would achieve stable equality this way. We tested it out on a few engines and they all retreated the knight. They thought the sacrifice was playable for white as well, but even when coached ahead a few moves, they did not like giving up a whole piece without concrete gains and evaluated the retreat higher.

While looking at the horrible positions for black that resulted, I realized that only a computer would be able to defend them, and that was the point. Computers love material and are incredible defenders. I was sure that Deep Blue would apply its fantastic defensive prowess to my position, evaluate it as fine for black, and therefore would decline to sacrifice the knight. I lost this bet, obviously, and spectacularly so, but the reason I lost it would not be clear for over a decade.

It may surprise you to find out that I was completely right in my evaluation of Deep Blue. It would never sacrifice the knight. And yet, it did. Why? Because of one of the most remarkable coincidences in the history of chess, or perhaps in history, period.

Here once more is Deep Blue coach Miguel Illescas in his 2009 interview, speaking about the fateful sixth game: "We were looking at all kinds of rubbish, such as 1.e4 a6 or 1.e4 b6, giving as many forced moves to the computer as we could. On this same morning we also introduced the move Knight takes e6 in the Caro-Kann, on the same day that Kasparov played it. That very morning we told Deep Blue, if Garry plays h6, take on e6 and don't check the database. Just play, don't think.... This was his bet, that the machine would never like this piece sacrifice for a pawn. And indeed, if we had given freedom to Deep Blue to choose, it would never have played it."

I will not repeat here the stream of profanities in Russian, English, and languages not yet invented that escaped my lips when I first read that paragraph. What in the hell was this? Two paragraphs after Illescas says IBM had hired Russian speakers to spy on me, he says the team entered this critical line into Deep Blue's book that morning? An obscure variation that I had only discussed with my team in the privacy of our suite at the Plaza Hotel that week in New York?

I'm no Nate Silver, but the odds of winning the lottery are quite attractive in comparison to those of the Deep Blue team entering a specific variation I had never played before in my life into the computer's book on the very same day it appeared on the board in the final game. And not only preparing the machine for the 4..Nd7 Caro-Kann—even during my brief dalliance with the Caro-Kann as a fifteen-year-old I played the 4..Bf5 line exclusively—but also forcing it to play 8.Nxe6 and doing this despite generally giving Deep Blue "a lot of freedom to play," in Illescas's own words.

Am I alone in failing to make the leap of faith required to believe that the timing of this could possibly be innocent? I am trying, but I am failing. The IBM team went to great lengths to ridicule me about my "hand of God" remarks, and maybe I deserved it. Deep Blue's moves were inexplicable, partly because IBM refused to explain them, but they were not human. But perhaps that was all part of the psychological warfare while other games were afoot. As Pynchon's Proverbs for Paranoids, number 3 says in *Gravity's Rainbow,* "If they can get you asking the wrong questions, they don't have to worry about the answers."

HAD I NOT melted down during game two and resigned prematurely, none of this would have mattered. Not only was the early resignation my fault, but allowing it to ruin my composure was the real fatal mistake. I played so far below my usual level after that that it has been a little embarrassing to go over the games for this book. As I said the day after the match on the Larry King show, rested and calmer, "I do not blame IBM, I blame myself." I then again challenged Deep Blue to the rematch I believed I deserved after winning the first and losing the second. I wanted to play under neutral conditions, and I wanted to see if I could beat it playing normal chess. Not anti-computer chess; Kasparov chess.

Of course this never happened. Deep Blue never played another game. I can sympathize a little with those who say that IBM had gotten what it wanted already, a giant PR boost and an increase in its stock value of $11.4 billion in just over one week. If the entire project cost IBM the estimated $20 million they said, it's an enviable return on investment even if only a fraction of those billions was due to the match. A loss to me in a rematch would be embarrassing, and, even if they won again, nobody remembers the second man to scale Mount Everest.

Later that night, I shared the elevator at the Plaza with the actor Charles Bronson. After a flicker of mutual recognition, he said, "Tough luck, man!" I said, "Yes, I'll try to do better next time." He shook his head and replied, "They'll never give you the chance." He was right.

A few days after the match, a Wall Street friend arranged a phone call between me and IBM CEO Lou Gerstner. I told him that since I had given his machine a rematch, he owed me and the world a rubber match. He was friendly and talked about the great potential, but I could tell it was never going to happen. It was polite, but it was a polite brush-off. He wasn't interested and, if Gerstner wasn't interested, IBM wasn't interested.

The argument that IBM abandoned Deep Blue and chess because I was mean to them at a press conference is a little odd. If that was their excuse for not participating in a rubber match, all right, but why take Deep Blue apart? "It's already directing traffic in Pittsburgh," wrote one columnist. Why not let it play in tournaments, or analyze games? Why not put it on the Internet to let millions of chess fans challenge it? Deep Blue was the biggest thing to come out of IBM in ages, so why shut it down overnight instead of capitalizing on a machine that enjoyed better name recognition than sports stars like Pete Sampras? If IBM was offended by my "wild allegations" about the integrity of Deep Blue's capabilities, then shutting it down immediately and limiting the team's ability to speak about it was a strange way to respond to them. Even a single game against anyone else would mean lowering Deep Blue from its pedestal, exposing it to scrutiny and criticism. It beat the champion and retired, Fischer-like, becoming as much myth as machine.

Chess fans and especially the computer chess community were outraged. They called it a crime against science, against the spirit of the quest for the holy grail started by Alan Turing and Claude Shannon. As Frederic Friedel put it to the *New York Times*, perhaps poking fun at Monty Newborn's comparison of Deep Blue's win to the moon landing,

"Deep Blue's victory over Kasparov was a milestone in artificial intelligence, but it's a crime that IBM didn't let it play again. It's like going to the moon and returning home without looking around."

As this book was headed to press in December 2016, my coauthor Mig Greengard exchanged emails with two members of the Deep Blue team, Murray Campbell and Joel Benjamin, and they kindly shared several items of interest. Campbell is still working on AI at IBM Research, and is still a chess enthusiast. As such, he says he would like to have seen a third match, and that they had already been working on how to further improve Deep Blue. He corrected contemporary press reports with the surprising news that Deep Blue was kept online in their lab until, he writes, "It was finally powered down in 2001. Half was donated to the Smithsonian (in 2002) and the other half to the Computer History Museum (in 2005). . . . It was still a respectable supercomputer. We didn't routinely run the chess hardware on the full system." More is the pity then, that it was hidden away from an inquiring public. Campbell also told Mig that his favorite part of his decades in computer chess (starting as a student in the late 1970s) was not the 1997 rematch itself, but the preparation for it, because the stress level was so high during the match. If only that had affected Deep Blue's play as it did mine!

GM Benjamin wrote to contradict his colleague Miguel Illescas's published recollections about game six, saying that it was he (Benjamin) who entered the fateful 8.Nxe6 move into Deep Blue's opening book, "a month or so before the match." That is, not "that very morning" of game six, as emphasized by Illescas with such vigor that this revelation was the headline of his interview. Benjamin said that he didn't dissent when the interview, and my incredulous response, were published in 2009 due to not wanting to publicly contradict his old teammate. This dispute between twelve-year-old and twenty-year-old human memories is another reason that all of Deep Blue's files and logs should have been released at the time, especially if it was never to play chess again. By dismantling Deep Blue, IBM killed the only objective witness.

As for me, I moved on. The world still needed a human world chess champion after all, it turned out. I was very disappointed that I was never going to have a chance to get my revenge on Deep Blue. And it was always in the back of my mind that we never got to reproduce all of Deep Blue's moves for posterity. It was an inverted Agatha Christie whodunit. There was ample circumstantial evidence and no shortage of motives, but it wasn't clear that there had ever been a crime.

I have been asked, "Did Deep Blue cheat?" more times than I could possibly count, and my honest answer has always been "I don't know." After twenty years of soul-searching, revelations, and analysis, my answer is now "no." As for IBM, the lengths they went to to win were a betrayal of fair competition, but the real victim of this betrayal was science.

CHAPTER 11

HUMAN PLUS MACHINE

OF THE MANY ATTEMPTS to make me, and humanity, feel better about my loss to Deep Blue, the only effective one was that it was also a win for humans, since humans built the machine. I said as much in many post-match interviews congratulating the team. Despite how ugly things had become in the rematch, I still felt like I was part of a grand experiment, even if I was in denial for a few years that it was ending.

Saying in reality, we all won because we're all human didn't actually make me feel better, but I've always been an optimist, and this was a reassuring, optimistic thing to say while facing years of the same questions about one of the most agonizing experiences of my life. I always wondered, if doing the same thing and expecting a different result is a form of insanity, what is asking the same question and expecting a different answer?

As for humanity, it recovered as quickly as it always does. Despite all the hype around the match and its potential implications for life on Earth, the world was not a different place on May 12, 1997, the day after the rematch, unless you were a world chess champion, a member of the Deep Blue team, or a programmer hoping to build the first machine to beat one. It's a little ironic that while after losing I went back to my day job, so to speak, the Deep Blue team had engineered their own obsolescence by defeating me.

Deep Blue's lack of a purpose beyond its narrow goal of beating me was proof of what AI advocates in the computer chess community and beyond had been warning about for years: that there would be precious little to learn from its victory beyond what we already knew was inevitable, that smarter programs on faster machines would beat the human world champion sometime around the year 2000. This isn't criticism, only a fact. The public mystique around chess had eroded at roughly the same pace as the public's ignorance of computers. Flashy headlines aside, as machines became more powerful and more common, the idea that a human could beat one at chess, and that it mattered, seemed increasingly outlandish.

Igor Aleksander, a British AI and neural networks pioneer, explained in his 2000 book, *How to Build a Mind*, "By the mid-1990s the number of people with some experience of using computers was many orders of magnitude greater than in the 1960s. In the Kasparov defeat they recognized that here was a great triumph for programmers, but not one that may compete with the human intelligence that helps us to lead our lives."

This did not mean that super-strong chess machines did not have an impact, only that their impact was limited to the chess world. The good news is that what happens in the chess world is frequently a useful preview for the rest of the world. I will look at three broad categories where, for better and for worse, my beloved game and I have been on the cutting edge of the rapidly changing relationship between humans and machines. As the curtain fell on a decade of human versus machine competition, it was time for human plus machine collaboration to take center stage. To put it more succinctly, if you can't beat 'em, join 'em.

The phrase "human plus machine" can apply to any use of technology since early man bashed something with a rock. Our progress in demonstrating our superiority over other animals is based not primarily on language, but on our creation and use of tools. The mental capacity to make things that improved survival chances led to the natural selection of better and better tool makers and tool users. It's true that many animals use objects as tools, from apes to crows to wasps, but there is a giant leap from picking up an object to use as a tool and visualizing the right instrument for a task and creating it.

Nearly everything a modern human does involves the use of technology. The shift in recent decades has been in how much our technology can do without us. Automation has steadily moved up the ladder of emulating and surpassing human capabilities, from heavy lifting to fine motor skills on the physical side to calculation and data analysis on the intellectual side. Machines are now moving seamlessly into supplementing fundamental cognitive functions like memory, as we let go of doing things that are more easily done by our computers and phones. Even before the iPhone turned smartphones into a standard accessory, the substitution effect our tech was having on our brains was an important topic.

The tech writer and journalist Cory Doctorow coined the term "outboard brain" for his blogging at the website Boing Boing in 2002. He wrote that it had "not only given me a central repository of all of the fruits of my labors in the information fields, but it also has increased the volume and quality of the yield. I know more, find more, and understand better than I ever have." Even if you don't blog, anyone who has ever searched through their own email or social media can identify with this feeling. Scrolling back through years of emails or Facebook posts is a far richer mnemonic than flipping through an old photo album. It's a chaotic, ad hoc diary that also includes contributions from friends and family.

A 2007 article in *Wired* updated this concept for the mobile era in an article titled "Your Outboard Brain Knows All." The first iPhone had only been available for a few months at the time, so the phenomenon author Clive Thompson described was about to become exponentially more powerful. He writes about BlackBerrys and Gmail and how there was little point in remembering people's phone numbers, or even your own, since phones "can store 500 numbers in their memory." He goes on to say, "The cyborg future is here. Almost without noticing it, we've outsourced important peripheral brain functions to the silicon around us."

This isn't so much revolutionary as it is another demonstration of the democratizing power of technology. Executives and other elites have been outsourcing many of their mundane cognitive functions to their secretaries and personal assistants forever. They used, and many still use, daily calendars and Rolodexes to organize and store the contact and schedule information that we now all carry on tiny computers in our pockets. Smartphones have made this process more powerful and efficient. We can look up anything now, not just phone numbers. We don't only look for a restaurant the way we would in an old phone book; we get restaurant recommendations from an algorithm and our phone can make a reservation or order delivery for us with just a few commands.

Following in the grand tradition of nearly every new technology, nobody started to panic about the potential downsides of cognitive outsourcing until kids starting doing it, and doing it in ways that their parents didn't understand. They type with their thumbs in ugly slang and funny symbols. They have short attention spans. They can't remember their own phone numbers. They spend more time on social media than they did with their friends irl (that's "in real life," my daughter tells me). They are becoming zombies, robbed of ambition and free will! *New York Times* columnist David Brooks reacted to the *Wired* article with a droll account of how he was giving in to the outsourced brain. "I had thought that the magic of the information age was that it allowed us to know more, but then I realized the magic of the information age is that it allows us to know less. . . ." Continuing, "You may wonder if in the process of outsourcing my thinking I am losing my individuality. Not so. . . . It's merely my autonomy that I'm losing."

A decade later, does anyone regret not having to memorize phone numbers or maps? Probably, yes, but they are the same type of people who lamented the lack of flaws in cloth and glass that wasn't produced by artisans, and who miss the hiss and pop of vinyl records. Do not confuse nostalgia with the loss of our humanity. Have we lost our free will to GPS devices and Amazon recommendations and personalized news feeds? Losing the serendipity of getting lost on an old country road, browsing through a bookstore, or thumbing through the printed newspaper can result in a marginal loss of well roundedness, I'm sure. But no one is preventing us from doing those things, especially since we have much more time in which to do them when satisfying our specific needs and wants has become so much easier.

We haven't lost free will; we have gained time that we don't yet know what to do with. We have gained incredible powers, virtual omniscience, but still lack the sense of purpose to apply them in ways that satisfy us. We have taken more steps in the advance of civilization, toward reducing the level of randomness and inefficiency in our lives. It's different, yes, and different can be disconcerting when it happens quickly, but that doesn't make it harmful. All this mockery and alarm will disappear soon after a member of the generation that grew up with smartphones gets a column in the *New York Times*.

Are there downsides to all of this mental outsourcing? Are we putting parts of our brain out of work by sending their cognitive processes across the border to our phones? Thompson wondered, "I'm a veritable genius when I'm on the grid, but am I mentally crippled when I'm not? Does an overreliance on machine memory shut down other important ways of understanding the world?" It's an important question, and by no means a new one. The acquisition of knowledge cannot not serve only to perform immediate tasks or answer a question, at least not if we wish to approach the higher goal of wisdom. Your phone can make you an instant expert on anything, thanks to Google and Wikipedia, and this is incredibly useful. Doing so doesn't make us dumber any more than encyclopedias, phone books, or librarians made us dumber. It is only the next stage of how our technology allows us to create and to interact with more information faster and faster—and it won't be the last stage. The danger isn't intellectual stagnation or an addiction to instant fact-finding missions. The real risk is substituting superficial knowledge for the type of understanding and insight that is required to create new things.

Expertise does not necessarily translate into applicable understanding, let alone wisdom. This debate has its origins with Socrates, and the thread continues across the usual route of Aristotle's *Nicomachean Ethics* and Descartes's *Principles of Philosophy*. What is wisdom? Is it accumulated knowledge? Humility in accepting our own ignorance? Knowing how to live well? Using our machines to acquire and retain more knowledge cannot be a bad thing on its own. The question is whether or not there is a type of cognitive opportunity cost. Having seen this entire process in action in a relatively quantifiable way thanks to chess, I think this is undeniable, but also that it isn't necessarily negative if we are aware of it. I reject the notion that everything must be a zero-sum game in which for every cognitive gain there is a corresponding loss. Big changes in how we manage our minds can, and often do, result in net positives. As with other aspects of what I call upgrading our mental software, self-awareness is the vital ingredient.

I've already mentioned how having a Grandmaster-strength computer in your home or pocket has encouraged the appearance of strong players around the world. It didn't only affect *who* plays chess, however. Chess machines have also had an impact on *how* human chess is played.

This doesn't refer to playing on the Internet or against computers, although that is also true. I'm referring to the way human Grandmasters play against each other after having spent their lives working with super-strong engines. It used to be that young players might acquire the style of their early coaches. If you worked with a coach who preferred sharp openings and speculative attacking play himself, it would influence his pupils to play similarly. You can make the identical case for tennis coaches and the teachers of fiction classes, I'm sure.

What happens when the early influential coach is a computer? The machine doesn't care about style or patterns or hundreds of years of established theory. It counts up the values of the chess pieces, analyzes a few billion moves, and counts them up again. It is entirely free of prejudice and doctrine, although it's true some programs are a little more aggressive or conservative depending on how their evaluations are tuned. The heavy use of computers for practice and analysis has contributed to the development of a generation of players who are almost as free of dogma as the machines with which they train.

Increasingly, a move isn't good or bad because it looks that way or because it hasn't been done that way before. It's simply good if it works and bad if it doesn't. Although we still require a strong measure of intuition, guidelines, and logic to play well, humans today are starting to play chess more like computers.

The talented kids I have been working with as part of the Kasparov Chess Foundation's Young Stars program for a decade are between eight and eighteen years old. They have all worked with strong chess machines since they first learned the moves and there is no doubt they have developed differently than the kids I worked with in the 1980s at the Botvinnik School in the Soviet Union. Because I am literally and figuratively "old school" myself, it is hard for me not to be critical of how these youngsters approach the game and of their lack of structured, dogmatic chess thinking. I also realize that you cannot argue with results, and that there are advantages as well as drawbacks to learning without so much dogma. Being able to explain why a move is good or bad in theory is not the same as being able to demonstrate it in practice.

The problem comes when the database and the engine go from coach to oracle. It happens quite often that I will ask one of the students about a move from one of their games, and why he made it. If the move comes early on, the answer is almost always, "Because that's the main line." That is, that's the theoretical move in the database, likely played by many Grandmasters before. Sometimes the move isn't theory, but the student prepared it with the help of an engine, so the answer is similar: "It's the best move." Maybe, yes, but, I always ask, why is it the best move? Why did all those Grandmasters play it? Why does the computer recommend it?

Then we often have a problem. Why? Because it is good. Why is it good? Answering that can take a lot of understanding and a lot of research. The openings have developed empirically over decades, sometimes over a century. If the bishop going to a certain square on move twelve in a specific variation is considered best, there's a whole story leading up to that moment, dozens or hundreds of games of trial and error that went into establishing why that move precisely now.

The kids want to skip all that and just start at the good part, where the previous analysis and old games tell them to go, before thinking for themselves. If you've been paying attention, you'll remember that's exactly how machines play, by using an opening book, a database of Grandmaster games and theory. Humans playing this way have the same drawbacks. What if there's an error in the book? What if you're following along blindly and your opponent has prepared a nasty novelty down the line you're following?

There is also a pragmatic logic to it, of course. If a move has been recommended by strong players and computers for a long time, it most likely is the best move. But unlike computers, humans run into two issues with accepting the database's verdict blindly. First, when you run out of memorized preparation you have to start using your brain. Even if you know that the position you have arrived at is supposed to be a good one for you, unless you've done some more substantial preparation you may have no idea what to do next. It's like taking a boat out into the middle of a lake and only realizing when the boat springs a leak that you don't know how to swim.

And what if your opponent diverges from the main line you memorized so assiduously? Computers don't care. They just pluck the right move out of the database if it's there and start thinking if it's not. But unless you have a good understanding of the general position, you might be in more trouble than your opponent even if his move isn't the best one according to the database. This is also why it's important to use your own brain while preparing, not just the engine. The machine will tell you what it thinks the best move is for both sides, not what the most likely response is, or the one that is the most difficult for the other player to handle. Overreliance on the machine can weaken, not enhance, your own comprehension if you take it at its word all the time. I tell my students that they have to use the engine to challenge their own preparation and analysis, not do it for them. It's not enough to know the best moves; you must also know why those moves are the best.

THE SECOND ISSUE is a deeper one that gets to the heart of how human-machine collaboration can help us be more creative, or less, depending on how we use our digital tools. Databases are composed not only of opening lines, but entire games. Although it happens sometimes, it's rare for two players to duplicate an entire game move for move. Even if both players know the game they are following, someone will eventually diverge in order to seek an advantage. That is, if two players are following a game that was lost by black, obviously the player with black needs to find an improvement somewhere along the way. The question then comes, where do you start looking for an improvement? At the spot where black blundered? That's a good start, and maybe you will avoid a disaster and have a decent result in the game if you improve there.

But when it comes to big innovations, you have to start earlier, not where the database ends. You have to dig into the tree of established moves that everyone assumes are the best because they've been played so many times before. This is one way I kept pressure on my opponents, year in and year out. They knew I was always working on deep opening improvements in popular lines, just like they were, but that I would also appear with new ideas very early on that occasionally led to a renaissance of a discarded opening or variation. This wasn't only good for my results, it was good for my feeling of creativity in general, not just in chess.

It's fine, especially for younger players, to rest on the shoulders of giants and imitate the top players in the openings, relying on them (and their computers) not to have made mistakes. This is the equivalent of the electronics companies whose products imitate those of the big brands, only at a lower price or with an extra feature or two tossed in. They don't create anything or innovate in any fundamental way. They are imitators only, and they compete with other imitators in a race to see who can copy faster and better. They can go out of business quickly when a new market opens up with even cheaper labor, or more efficient manufacturing, unless they learn to innovate themselves.

And so it is with chess thinking, business thinking, and with pursuing innovation in general. The earlier on in the development tree you look, the bigger the potential for disruption is, and the more work it will take to achieve. If we only rely on our machines to show us how to be good imitators, we will never take that next step to becoming creative innovators. The world is a big enough place for every kind of success, of course. Some might argue that Apple, for example, has left its disruptive roots and become little more than an imitator with great fashion sense and even better marketing, since they didn't create most of the technology inside the wildly popular products they make. Not every great singer writes her own songs, however, and Apple's shareholders and consumers clearly believe that their design and brand add a lot of value to their products. But if everyone imitates, soon there will be nothing new to imitate. Demand can be stimulated by incremental product diversification for only so long.

The entrepreneur and venture capitalist Max Levchin used a good expression for this effect referring to Silicon Valley and tech start-ups, and I like it for just about everything. While we were working on a book project together a few years ago, he called it "innovating at the margins." That is, looking for small efficiencies instead of taking on more substantial risks in the main area of business. Levchin has been interested in online payment and alternative currencies since cofounding PayPal in 1998, and he described how most of these services are trying to squeeze nickels out of the 2 to 3 percent banking fees while leaving the principle risk to the big banks. That adds convenience and efficiency, but it's not disruption.

This is a shame, because the potential for change is much greater than our appetite for it. Our increasingly powerful machines give us the security to be more ambitious and better prepared, but we still have to make the choice to do it. Technology has lowered the barrier to entry in dozens of business sectors, which should prompt more experimentation and investment. Powerful models allow us to simulate the impact of change better than ever, lowering risk.

Once again using chess machines as our favorite *Drosophila* fruit-fly test subject and metaphor, Grandmasters have used the ability to prepare with engines and databases to play riskier, more experimental opening variations. Many members of the chess community were afraid that super-strong machines would damage professional chess irreparably by reducing Grandmasters to the role of puppets doing little more than relaying the moves their engines told them were the best. And, to be fair, there is an element of this at a level below elite, as there has always been a class of imitators below the innovators. But at the top level, the effect has been the opposite, with a few glaring exceptions.

With the safety net of using an engine at home during preparation, many GMs are more willing to play sharp variations over the board in tournaments. The lure of catching your opponent in a deadly piece of preparation is stronger than the chance of it backfiring. Human recall isn't perfect, and your opponent might also be well prepared, or come up with something you didn't think of at home. Either way, there are still plenty of exciting variations and games being played.

The exception is what might be called an anti-computer movement in elite human chess. It involves the rise of opening variations that are very positional and strategic and therefore not as vulnerable to a computer-created landmine found by your opponent. Principal among these is the Berlin Defense of the Ruy Lopez, which Kramnik used against me to great effect in our 2000 world championship match. The queens come off the board very early in the Berlin, and although white has his usual slight advantage, the positions require very subtle play of the kind even today's strong engines often find befuddling. While some players have been pushed in a creative direction by preparing with engines, others have been pushed in a more conservative direction by the threat of their opponents' engines, our own chessic Luddite movement. Unfortunately, to my taste, the Berlin Defense is currently the dominant strain. I say it's unfortunate not only because I personally find these positions tedious, which is why Kramnik wisely selected them. These subtle positions also tend toward equality and many drawn games, which makes chess less attractive for fans who like action on the board and more wins and losses than draws.

The availability of millions of games at one's fingertips in a database is also making the game's best players younger. That is, players are joining the elite at an earlier age than ever before. Bobby Fischer set the standard for decades by joining the elite at fourteen, when he won the US championship. He became a Grandmaster officially the next year, in 1958, although he was clearly already playing at GM level. This record stood for thirty-three years, when Hungary's Judit Polgár beat it by just a few months in 1991. She wouldn't own the record for long, however, only until 1994, and then the floodgates opened. Fischer's record has now been surpassed by no fewer than thirty players.

The record holder since 2002 has been Ukrainian-born Sergey Karjakin, who now plays for Russia. He achieved the title at the age of twelve years and seven months. He's definitely no Fischer, but in another indication that prodigy is often destiny, he reached the world championship final in November 2016, where he lost to incumbent champion Magnus Carlsen, who was also born in 1990. (Carlsen is number three on the "youngest GM ever" list, at thirteen years, four months.)

Finding the overwhelming correlation only requires a look at the date when all the records started to fall. Nineteen fifty-eight, 1991, 1994, and then the flood, 1997, 1999, 2000, and over two dozen more in the next decade, all becoming GMs earlier than Fischer. The beginning of the boom coincides exactly with the spread of professional training software with strong engines and online play.

There are a few peripheral factors in this incredible pace of production of young GMs, including how gaining the GM title as early as possible became a fashion, and how ratings have inflated over the years, making the bar of 2500 relatively easier to obtain, if still far from a trifle. Becoming a teenage GM used to signify generational talent; now it is practically routine. Fischer not only got the GM title at fifteen, he qualified for the world championship Candidates tournament, putting him among the world's eight best players. There are far more opportunities today for young would-be Grandmasters to acquire the title qualifications and the pace of events is far faster. I didn't get the official title until I was seventeen, despite having qualified for the Soviet championship at the age of fifteen, one of the strongest tournaments in the world. I was awarded the title at the FIDE Congress in Malta in December 1980, and on the January 1981 rating list I was ranked sixth in the world. One amusing indication of how difficult it used to be to obtain the Grandmaster title is a story I was told by Yasser Seirawan about Walter Browne. Browne, who passed away in 2015, was complaining about the proliferation of GM titles in the 1990s, when dozens were awarded at every congress. He said, "When I got my title in 1970 there were only two of us. The other guy was Karpov and they weren't so sure about him!"

Absorbing the thousands of essential patterns and opening moves required to approach the Grandmaster level used to take many years, a slow process indicative of Gladwell's "ten thousand hours to become an expert" I discussed earlier. Practice has shown that technology can greatly reduce that time by making training far more efficient. Today's teens, and increasingly preteens, accelerate the process by plugging into a digital fire hose of chess information and making full use of the superiority of the young mind to retain it all. Instead of ten thousand hours, it would be more accurate to say that ten thousand patterns are required, to pick an arbitrary number, or perhaps fifty thousand positions.

I worked with Carlsen for a year in 2009, as he was making his way up the chess Mount Olympus. He was clearly a generational talent, ranked fourth in the world already at just eighteen years of age. I noted that he was impressively judicious in his use of computer engines. He wasn't mesmerized, as many of my young students are, by the illusory perfection of machine analysis. Carlsen is comfortable with his own strengths and sees the machine appropriately, as a tool, not an oracle. This helps him in training because he builds up critical mental problem-solving muscles instead of simply having the answer handed to him by an engine. It also helps him at the board, because when he needs to solve a tough problem he doesn't mentally reach for the mouse.

Compare that to what you do when you can't remember something and reflexively reach for your phone. Do you at least pause for a minute to see if you can figure it out on your own? You may not be a world champion in training, and you might just be looking up some movie trivia or a friend's email address, but it is still worth getting those cognitive muscles a little exercise on occasion. Acquiring and remembering knowledge has value if we employ it creatively, the way the human brain is designed to work. It tosses all that trivia together and turns it into insight and ideas, often while we aren't even conscious of it. We may not wander through bookstores very often anymore, but we must still let our minds wander in search of inspiration.

The diversity of the geographic regions represented by those talented chess youngsters is also notable. The former-Soviet powerhouses are all there, but also India, Norway, China, Peru, and Vietnam. On a national scale, you can see the same effect in the United States. The American chess world used to center in New York City almost completely. The young stars the Kasparov Chess Foundation brings together also represent California, Wisconsin, Utah, Florida, Alabama, and Texas. For the last two decades, especially with the proliferation of the Internet and cell phones, a dominant topic has been how technology will enable people from all over the world to become entrepreneurs, or scientists, or anything they want despite where they live. Here, again, our little chess *Drosophila* has already shown the way. The talent is out there; people only need the tools to express it.

Chess sneaks through the cracks of cultural, geographic, technological, and economic barriers, disguised as an innocuous pastime. Again and again, it serves as a model for everything from artificial intelligence to online gaming to problem solving and gamification in education. The boom in young Grandmasters and how they think should serve as an example for traditional education as well, with similar cautions. Kids are capable of learning far more, far faster, than traditional education methods allow for. They are already doing it mostly on their own, living and playing in a far more complex environment than the one their parents grew up in.

I occasionally wonder if I'd have become a chess champion had my home and neighborhood in 1960s Baku possessed the countless diversions available to kids today. As does every generation of parents, I lament all the distractions pulling at the attention of my youngest kids. But this is their world, and we need to prepare them for it, not futilely attempt to shield them from it. Kids thrive on connections and creation and they can be empowered by today's technology to connect and create in limitless ways. The kids who go to schools that embrace this empowerment most ably will thrive.

That our classrooms still mostly look like they did a hundred years ago isn't quaint; it's absurd. How can a teacher or even a stack of books be the sole source of information for kids who can access the sum of all human knowledge in seconds from a device in their pockets, and do so far more quickly than their teachers or parents? The world is changing too quickly to teach kids everything they need to know; they must be given the methods and means to teach themselves. This means creative problem-solving, dynamic collaboration online and off, real-time research, and the ability to modify and make their own digital tools.

Despite the affluence and high level of technology in the United States, Western Europe, and in Asia's traditional economic leaders, the potential for rapid change in education is likely in the developing world. There is little reason for them to try to catch up to the developed world by imitating education methods that are becoming obsolete. Just like the people in many poorer nations have adopted smartphones and virtual currencies while skipping the steps of personal computers or traditional banking, they can adopt dynamic new education paradigms very quickly since there are fewer existing structures to replace.

They are aided by how far we have come in making powerful technology easily accessible. A room full of kids can assemble their own digital textbooks and syllabus in a few minutes of drag-and-drop on a tablet, collaborating from the very start. I know it's possible because I've seen it done with chess courses. The kids can access new material on demand and the instructors might be anywhere in the world, available 24/7 instead of only during school hours.

Wealthy nations approach education in the same way a wealthy aristocratic family approaches investing. The status quo has been good for a long time; why rock the boat? I've spoken at many education conferences in the past few years, from Paris to Jerusalem to New York, and I've never seen such a conservative mindset in any other sector. Not only the administrators and bureaucrats, but the teachers and parents as well. Everyone except for the kids. The prevailing attitude is that education is too important to take risks. My response is that education is too important *not* to take risks. We need to find out what works and the only way to do that is to experiment. The kids can handle it. They are already doing it on their own. It's the adults who are afraid.

MY MATCH against India's Viswanathan Anand in New York in 1995 was the first that saw the use of computer engines in championship preparation. My team of human seconds and I had decided we could incorporate Fritz 4 into our preparation routine if we used it only as a sort of fact-checking calculator. We didn't trust it with anything strategic, but it could be a real time-saver to work out extremely tactical positions without risking silly oversights.

Anand and I had drawn eight consecutive games to begin the match. I was favored to defend my title against "Vishy" at the start, but as our timid stretch of draws continued, pundits began to wonder if I'd lost my touch. And, honestly, I was getting a little worried myself. Anand was well prepared and I was playing without much confidence. A stretch of poor play can get you to start second-guessing your decisions, leading to more poor play. I would find inspiration not at the board, however, but in my team's apartment in lower Manhattan. I came up with an amazing piece sacrifice to use against Anand's preferred defense against the king's pawn, the Open Ruy Lopez. My team and I spent the entire weekend hammering out the incredibly complex tactics that followed the sacrifice, and here the machine was quite useful, even as relatively weak as engines were back then.

The problem was, I didn't have white in the next game. I was so anxious to play this wonderful idea that we had neglected to focus on the next game, in which I had black, not white. It felt like an annoyance to have to play another game before I would be able to spring this dazzling new idea on Anand and the world. The old barnyard proverbs about carts and horses and counting chickens quickly came home to roost when I lost terribly. This is not to take away from Anand's play, since he played the game very strongly and deserved the win. But I was kicking myself for becoming distracted and I knew I had to get my focus back so I wouldn't squander my new idea the next day. I was now behind the match at the halfway point.

The day finally came and I was crackling with energy. I hoped Anand could not somehow read my plans on my face. Had he diverged from the sixth game and played something other than the Open Ruy Lopez I would have been quite shaken. I was so wound up that when the arbiter accidentally dropped the clock on the board with a bang I jumped and covered my face for a moment.

To my great relief, Anand repeated his opening as I had hoped, and we followed along until move fourteen. From a certain perspective, it made sense for Anand to repeat. That game had gone well for him, so why not? It was up to me to find an improvement. On the other hand, did he imagine that I would repeat it had I not found a strong novelty? It showed his confidence in his opening preparation, which had been very good in the first half of the match. And no one could rightly expect what happened next.

I varied from game six on move fourteen with a bishop move that had been analyzed before by other players, but incompletely. Mikhail Tal, the world champion renowned for his astounding tactical vision, had proposed this sacrifice many years earlier, but his suggestion had been discarded because the follow-ups he gave were insufficient for white. Other analysts had also rated the idea as a spectacular dud. I had found an incredible twist that would turn the evaluation upside down, at least for one crucial game. Instead of bringing my knight to the center, as Tal recommended and as seemed logical, I brought it to the side. There the knight would protect my rook, attack its black counterpart, and not block my other pieces from joining in the attack on the black king.

After finally banging out the move that had dominated my thoughts for the past three days, I couldn't contain my nervous energy any longer and jumped up to go pace a little. I let the door of the playing area slam behind me, interpreted by some as a rude attempt at psychological warfare. It was only nerves, however, as I was more than happy to let my move speak for me in this case. The novelty offered the sacrifice of an entire rook in exchange for a vicious attack on Anand's king. We hadn't found a refutation in our preparation and Anand spent a remarkable forty-five minutes looking for an answer. (Especially remarkable for him, one of the fastest players in chess history.)

I was still in my deep preparation as Anand looked for a way out of the trap. He found the best defense in several tough spots, and I still had to play precisely to score my first win of the match and draw even. There were ten games still to play, but the initiative was very much on my side now. I sprung another surprise in the very next game, playing the Dragon Variation of the Sicilian for the first time in my life. I won that game, and then two out of the next three as well to take a big lead in the match that I never relinquished.

It's interesting to go back over these games and all the articles and books that were written about the match and compare it all to my matches with Deep Blue. So much of the writing from both my camp and Anand's, as well by journalists and analysts, focuses on psychology. Take this from American GM Patrick Wolff, one of Anand's seconds during the match, after that tenth game. "After game 9, all of us in Anand's camp were elated. After game 10, we were dejected. Such strong passions play an important role in a match. A match is not a test of one's absolute ability to play chess—whatever that is—but of how well one has played those particular games. Therefore, the ability to monitor and control one's mood is of great importance in determining the match outcome."

After eight straight draws to start the match followed by a win, Anand lost four of the next five, essentially ending the match with six games still remaining. Anand did not suddenly become a much worse player after game ten, nor did I become a much stronger one. And as much credit as I would like to give my team and me for our opening surprises, they weren't what made Anand play far below his usual level for that horrible stretch. He lost one game to a strong novelty, was faced with a surprising defense in the next game and blundered, and was never able to recover the composure required for consistency. In a way, it was lucky I hadn't found that strong novelty in my pre-match preparation. Had the fateful tenth game instead been the second game, he would have had time to recover.

This isn't a critique of Vishy Anand; it's a critique of humanness. A similar syndrome would envelope me eighteen months later in my rematch with Deep Blue, and being aware of it was useless in countering its effects. Our emotions rule over our cognition in countless ways, many of which we cannot explain. Some players actually seem to play better when they are on the ropes. They dig in and defend like tigers, taking it as a challenge to rise to. Viktor Korchnoi was like this; he enjoyed grabbing a pawn even if it meant weathering a brutal attack. A man who had survived the Siege of Leningrad as a boy was not going to be intimidated at the chessboard. This sort of mental robustness is rare, even among elite Grandmasters. Mistakes almost never walk alone.

This is just as true in every walk of life. Many studies have shown that depression, or a simple lack of self-confidence, results in decision making that is slower, more conservative, and inferior in quality. Pessimism leads to what the psychologists call "a heightened sense of potential disappointment in the expected outcome" of one's decisions. This leads to indecisiveness and the desire to avoid or postpone consequential decisions. If those afflicted employ typical decision-making techniques, their results barely suffer at all. The breakdown occurs earlier, with the depression interfering with the fundamental habits of making logical decisions.

Intuition is the product of experience and confidence. And here I mean "product" in the mathematical sense, as the equation *intuition = experience x confidence.* It is the ability to act reflexively on knowledge that has been deeply absorbed and understood. Depression short-circuits intuition by inhibiting the confidence required to turn that processed experience into action.

Emotional influence is only one of the many ways in which humans act irrationally and unpredictably. Economic theory is predicated on the fact that people are "rational actors," that we will always decide based on what is in our best interests. This is probably why economics is called the "dismal science" and why there is a saying that economists have as much effect on the economy as weather forecasters have on the weather. Humans often aren't rational at all, not in groups and not individually.

One of the simplest and most powerful examples of how vulnerable we are to false intuition is the "Monte Carlo fallacy," also called the "gambler's fallacy." Assuming a fair coin and a fair flip, if the coin comes up heads twenty times in row, what are the odds it will come up heads again on the next flip? Surely twenty-one consecutive heads is hugely against the odds. The instinct would be to bet on tails, assuming that some statistical regression will occur in your favor eventually.

This is entirely wrong, and the belief that it is true is a big part of why the gambling empires in Las Vegas and Macau don't have to worry about paying their immense electricity bills. Each flip has the same fifty-fifty odds, no matter how many come up in a row or any other order. Twenty-one consecutive heads is no more and no less likely than any other sequence of twenty-one flips.

Even if you've never heard of this fallacy, on some level you know this is accurate. You know that each flip is fifty-fifty and isn't influenced by whatever happened before. And yet . . . the instinct is very strong to think that the odds are somehow altered by previous events. The fallacy supposedly draws its name from an oft-repeated story about a roulette wheel in a Monte Carlo casino where the ball fell in black twenty-six times in a row. Incredibly rare, yes, but if you're paying attention you'd realize that it was no rarer than every other one of the 67,108,863 sequences possible after twenty-six spins. It was only more notable to pattern-obsessed humans, who lost, so the story goes, millions of francs betting that red was more likely.

You can see why computers have a certain advantage in games where streaks of lucky or unlucky cards or dice rolls can influence the decision making of humans. Machines don't look for patterns in randomness, or least if they're programmed to, they don't find any the way our minds often do.

The fascinating work of researchers like Daniel Kahneman, Amos Tversky, and Dan Ariely has demonstrated how terrible human beings can be at thinking logically. For all the immense power of the human mind, it is very easy to fool. I'm a firm believer in the power of human intuition and how we must cultivate it by relying on it, but I cannot deny that my faith has been shaken by reading books like Kahneman's *Thinking, Fast and Slow* and Ariely's *Predictably Irrational*. After reading their work, you might wonder how we survive at all.

Just like chess Grandmasters do at the board, we rely on assumptions and heuristics to make sense of the complexity around us. We do not calculate every decision by brute force, checking every possible outcome. It is inefficient and unnecessary to do so, because generally we get by pretty well with our assumptions. But when they are isolated by researchers, or exploited by advertisers, politicians, and other con artists, you can see how we could all use a little objective oversight, which is where our machines can help us. Not merely by providing the right answers, but by showing us how idiosyncratic and easily influenced our thinking can be. Becoming aware of these fallacies and cognitive blind spots won't prevent them entirely, but it's a big step toward combating them.

During my annual visit to Oxford in 2015, I gave a seminar on decision making to a group of students at the Saïd Business School. For one segment, I performed an experiment based on those described by Daniel Kahneman to test what cognitive psychologists call the "anchoring effect" in our decision making. Would it work on a group of MBA students even though they knew I was trying to trick them?

I broke them into seven groups of five or six students each, and each group got a slightly different version of a handout containing six questions. The first three questions were all variations on the following yes-or-no questions:

> Was Gandhi more or less than 25 years old when he died?
>
> Is the tallest tree in the world taller or shorter than 60 feet (18 meters)?
>
> Is the average annual temperature in Damascus above or below 3°C (37°F)?

The next three questions were the same for all seven groups:

> How old was Gandhi when he died?
>
> How tall is the tallest tree in the world?
>
> What is the average annual temperature in Damascus?

The handouts varied only in the numbers provided in the first three questions. They got higher by roughly 25 percent for each group. That is, for group two the Gandhi question was "more or less than 30 years old," the tree was "taller or shorter than 100 feet (31 meters)," and the

Damascus temperature was "above or below 8°C (46°F)," etc. By the time the handouts reached group seven, the numbers were 125 years, 1300 feet (400 meters), and 48°C (118°F).

I tried to pick things that people likely wouldn't know for sure, but would have a strong intuition about. Everyone knows Gandhi wasn't under twenty-five or over one hundred and twenty-five when he died, and that the tallest tree in the world must be much taller than sixty feet. The point wasn't the first three questions, however. They were only there to influence the students' answers to questions four through six, and they certainly did. Keep in mind that no actual information was provided to the students on the handout, only questions, and that they had been warned to think objectively because I was trying to trick them.

The averages of the students' answers in group one were 72 years, 30 meters, and 11.4°C. In group five, they were 78 years, 112 meters, and 24°C. In group seven, 79 years, 136 meters, and 31.2°C. With only two exceptions, the numbers got higher on every question in every group's averages. (One group had three students from India who knew exactly how old Gandhi was when he died, 78. The other answers are 379.7 feet [115.7 meters] and 11.2°C [52.1°F].) The temperature averages went 11.4, 18.1, 21.3, 21,8, 24, 30.7, 31.2. The numbers in the first set of questions directly impacted the students' answers in the second set despite not imparting any useful knowledge and despite being obviously exaggerated in some cases.

Kahneman describes this anchoring effect with even weaker influences, such as spinning a wheel of random numbers in front of a class before having them answer questions about the values of things. As you can guess, the higher the number that came up on the wheel, the higher the average of the students' estimations. Even if the students are told to ignore the wheel, the averages follow the number. Your brain can be very powerful at tricking itself.

We suffer from similar irrationalities and cognitive delusions at the chessboard as we do in life. We often make impulsive moves when careful analysis refutes our plans. We fall in love with our plans and refuse to admit new evidence against them. We allow confirmation bias to influence us into thinking what we believe is right, despite what the data may say. We trick ourselves into seeing patterns in randomness and correlations where none exist.

In chess analysis, having an engine peeking over your shoulder while you work is very useful, but it can also enslave you and intimidate you if it's on all the time. Unless you've got Pocket Fritz in your pocket, you won't have that help available when you're playing a game. And while using your phone isn't cheating in real life, you might develop a cognitive limp from an overreliance on a digital crutch. The goal must be to use these powerful and objective tools not only to do better analysis and make better decisions in the moment, but also to make us into better decision makers.

Every chess move I made in my career represented a decision. Because of the circumscribed nature of the game of chess, each of these decisions could be analyzed and evaluated for quality. Life is not so clear-cut and our day-to-day decision making isn't as susceptible to objective analysis as chess moves. But this is changing. Our machines are increasingly capable of helping us become more aware of our decisions as we feed them more and more data about our lives. Your personal finance is tracked online by banks and brokers as well as by specialized sites and apps. Education goals can be monitored and performance tracked. Your health is monitored by a device on your wrist and apps for counting calories and counting sit-ups. Studies tell us that we consistently overestimate how much exercise we get and underestimate how much we eat. Why? It serves our ends of thinking well of ourselves, and of eating more snacks. Human plus machine can keep you honest, as long as you are honest with your machines.

We can use all of these tools to test our own assumptions and decisions, part of that mental muscle building I mentioned earlier. How long do you think it will take to complete a project or to achieve some other goal? Then go back and see how accurate your estimation was. If it was way off, why was it wrong? Checklists and goalposts are vital to disciplined thinking and strategic planning. We often stop doing these things outside of a rigid work environment, but they are very useful and today's digital tools make them very easy to maintain.

I am often described as being a very impulsive person, and I don't disagree, which would seem to be a drawback for a chess champion. I've been asked more than once how I reconcile my "jump first, ask questions later" attitude with the cool objectivity required to play elite chess. I always answer first off that I have no universal tips or tricks for becoming a disciplined thinker. We are all different, and what works for me might not work for others. I was blessed with a devoted mother and a great teacher who focused on discipline from the beginning instead of indulging my impulsive nature. Klara Kasparova and Mikhail Botvinnik both understood that my talent would not be crushed or disappear by their attempts to rein in my impetuousness.

My next answer is that you have to be brutally honest where it counts the most. I tried to be as objective as any machine when going over my games. If I wasn't always completely successful, I would say I was successful enough. If you are truthful and diligent with collecting data and making your evaluations, you will find you get better and better at making correct estimations.

Like my chess students who use engines to train themselves into making more objective, accurate decisions, you can use our increasingly intelligent machines to become a better decision maker, not only by outsourcing some of those decisions, but by observing and analyzing the ones you make more objectively. All the data in the world won't help you overcome your biases if you don't listen to it. Stop making excuses and rationalizations that are only your mind tricking you into doing what it wants to do. It can be hard to let the data speak for itself. After all, we aren't machines.

If you remember Moravec's paradox, it says that what machines are good at is where humans are weak, and vice versa. This is well illustrated by chess, and this gave me an idea for an experiment. What if, instead of human versus machine, we played as partners? My brainchild saw the light of day in a match in 1998 in León, Spain, and we called it Advanced Chess. Each player had a PC at hand running the chess software of his choice during the game. The idea was to create the highest level of chess ever played, a synthesis of the best of man and machine.

I wasn't aware of it at the time, but the great British AI researcher and game theorist Donald Michie had proposed this concept as early as 1972, in an article on machine chess in *New Scientist* magazine. He called it "consultation chess" and thought it would be interesting to see how much a human player would improve by being able to access a "brute force component" during a game. Engines wouldn't have been of much use in 1972, however, so despite Michie's suggestion and a few others over the years, it was never put to the test.

Although I had prepared for the unusual format, my León match against the Bulgarian Veselin Topalov, one of the top players at the time, was full of strange sensations. Having an engine available during play was as disquieting as it was exciting. Being able to access a database of a few million games also meant that we didn't have to strain our memories nearly as much in the opening. But since we both had equal access to the same database, the advantage still came down to creating a new idea at some point.

Having a computer partner also meant never having to worry about making a tactical blunder. The computer could project the consequences of each move we considered, pointing out possible outcomes and countermoves we might otherwise have missed. With that taken care of for us, we could concentrate on strategic planning instead of spending so much time on laborious calculations. Human creativity was even more paramount under these conditions, not less.

Despite access to the best of both worlds, my games with Topalov were far from the perfection I sought. We were playing on the clock and had limited time to consult with our silicon assistants. Still, the results were notable. A month earlier I had defeated the Bulgarian in a match of regular rapid chess 4–0. In contrast, our Advanced Chess match ended in a 3–3 draw. My advantage in calculating tactics had been nullified by the machine.

León continued to host Advanced Chess events for years, often producing interesting insights. One of the things I liked about it was that the players' computer screens could be mirrored for the audience. It was like having a hidden camera inside a Grandmaster's mind as they looked at different variations. Even without engine assistance, this sort of real-time exposition of a player's thinking is very interesting. The entire analysis tree produced by each player during the game can be preserved and compared afterward with that of the other player to see how differently they approached key positions.

Even more notable was how the advanced chess experiment continued. In 2005, the online chess-playing site Playchess hosted what it called a "freestyle" tournament in which anyone could compete in teams with other players or computers. Normally, "anti-cheating" algorithms are employed by online sites to prevent, or at least discourage, players from cheating with computer assistance. (I wonder if these detection algorithms, which employ diagnostic analysis of moves and calculate probabilities, are any less "intelligent" than the playing programs they detect.)

Lured by the substantial prize money, several groups of strong Grandmasters working with several computers at the same time entered the competition. At first, the results seemed predictable. The teams of human plus machine dominated even the strongest computers. The chess machine Hydra, which was a chess-specific supercomputer like Deep Blue based in the United Arab Emirates, was no match for a strong human player using an ordinary computer. Human strategic guidance combined with the tactical acuity of a computer was overwhelming.

The surprise came at the conclusion of the event. The winner was revealed to be not a Grandmaster with a state-of-the-art PC, but a pair of amateur American players, Steven Cramton and Zackary Stephen, using three computers at the same time. Their skill at manipulating and "coaching" their computers to look very deeply into positions effectively counteracted the superior chess understanding of their Grandmaster opponents and the greater computational power of other participants. It was a triumph of process. A clever process beat superior knowledge and superior technology. It didn't render knowledge and technology obsolete, of course, but it illustrated the power of efficiency and coordination to dramatically improve results. I represented my conclusion like this: *weak human + machine + better process* was superior to a strong computer alone and, more remarkably, superior to a *strong human + machine + inferior process.*

I wrote about the freestyle chess result and my conclusion in *How Life Imitates Chess* and expanded on it a little in a 2010 article for the *New York Review of Books*. The response it received was quite a surprise, as calls and emails came in from all over the world about my little formulation. Invitations to lecture about the importance of superior process in human-machine collaboration came in from Google and other Silicon Valley companies as well as investment firms and business software companies who told me that they had been trying to make this case to potential customers for years. Alan Trefler, the founder and CEO of Pegasystems in Cambridge, Massachusetts, was a serious chess player and chess programmer in his youth. Pega makes business process management software and Trefler was quite excited about my article. "That's exactly what we do, I've just never been able to explain it that well!"

It's a little amusing to see versions of it referred to now as "Kasparov's law," although I guess we don't usually get to decide such things for ourselves. Timing had as much to do with the article's success as anything else. Intelligent machines have been making great advances thanks to machine learning and other techniques, but in many cases they are reaching the practical limits of data-based intelligence. Going from a few thousand examples to a few billion examples makes a big difference. Going from a few billion to a few trillion may not. In response, in an ironic twist after decades of trying to replace human intelligence with algorithms, the goal of many companies and researchers now is how to get the human mind back into the process of analyzing and deciding in an ocean of data. As with chess programs, which went from knowledge to brute force and then had to tilt back a bit toward knowledge as brute force ran into diminishing returns. The key again is the process, because that is something that only humans can design.

The interface remains one barrier to collaborating efficiently. Humans do many things better than machines, from visual recognition to interpreting meaning, but how to get the humans and machines working together in a way that makes the most of the strength of each without slowing the computer to a crawl? IBM is one of many companies now focusing on "IA," or intelligence amplification, to use information technology as a tool to enhance human decisions instead of replacing them with autonomous AI systems. Once again, our kids are way ahead of us. They prefer photos to symbols, symbols to texts, texts to emails, and emails to voicemails. It's all about speed. They are working out ways of communicating faster with each other and with their devices.

A line of code, a mouse, a finger, a voice command, these are all primitive analog tools compared to the incredible capacity of our machines today. We need a new generation of intelligent tools that will perform as human-machine (and machine-human) interpreters. Groups of people speaking together in a meeting is fine since everyone is operating at human speed. But now that machines are entering the decision-making space, how do we interact with them? Many jobs will continue to be lost to intelligent automation, but if you're looking for a field that will be booming for many years, get into human-machine collaboration and process architecture and design. This isn't just "UX," user experience, but entirely new ways of bringing human-machine coordination into diverse fields and creating the new tools we need in order to do so.

Our algorithms will continue to get smarter and our hardware faster. Machines gradually improve at a given task to the point where they no longer benefit from human partnership, the way elevators outgrew their operators. This is the way it goes, and will continue to go if we are lucky enough to enjoy a continued stream of technological advances. I assume we will, and this is very good news because the alternative is stagnation and declining living standards. To keep ahead of the machines, we must not try to slow them down because that slows us down as well. We must speed them up. We must give them, and ourselves, plenty of room to grow. We must go forward, outward, and upward.

CONCLUSION

ONWARD AND UPWARD

IN 1958, American science fiction legend Isaac Asimov wrote a very short story called "The Feeling of Power." In it, lowly technician Myron Aub discovers that he is capable of duplicating the work of his computer by multiplying two numbers together on a piece of paper. Amazing! This miraculous discovery makes its way up the chain of command, where the generals and politicians are stunned by Aub's black magic. The top general is intrigued by the possibility that human calculations could give Earth's forces a decisive advantage against those of planet Deneb, long locked in a stalemate of computer-controlled maneuvers.

Aub's remarkable ability to do math on paper and even in his head, nicknamed "graphitics," travels all the way up the ranks to the president, who is excited by the potential after this pitch by a congressman: "We will combine the mechanics of computation with human thought; we will have the equivalent of intelligent computers; billions of them. I can't predict what the consequences will be in detail but they will be incalculable. . . . In theory there is nothing the computer can do that the human mind cannot do. The computer merely takes a finite amount of data and performs a finite number of operations upon them. The human mind can duplicate the process."

The president is thusly convinced to launch Project Number in order to explore the military possibilities. The conclusion is typically wry Asimov. The general tells the assembled team, including the newly promoted Aub, that his vision is to replace the expensive computers on spaceships and missiles with men using graphitics. He concludes, "The exigencies of war compel us to remember one thing. A man is much more dispensable than a computer." This is too much for poor Aub, who goes back to his room and kills himself, leaving a note saying he couldn't face the responsibility of having invented graphitics, that he had hoped it would be put to use for the good of mankind.

Asimov was fascinated by how human–machine relationships would evolve, as best evidenced by his more famous stories about robots. And based on the publication date of "The Feeling of Power," it's certain that Asimov had more than a parody of human stupefaction and replacement by machines on his mind. The hydrogen bomb had recently been tested by both the US and the USSR, and the promise of nuclear fusion power was being debated against the possibility of a world-ending catastrophe. Would our vast new powers be used for good, or for destruction?

For most of human history, the answer has been both, although we have taken great strides in the past few decades of doing far more good than harm. Despite what you may think after watching an hour of cable news, we lead healthier, longer, and safer lives today than at any time in human history. My last book, *Winter Is Coming*, warned that this was a geopolitical trend, a season, and that it was reversible if we did not take action to preserve it. Our technology is not concerned about good or evil. It is agnostic. The same smartphone that brings people together all over the world can be used to connect with family or to plan a terrorist attack. The ethics are in how we humans use it, not whether or not we should build it.

There are many happily contradictory threads in this discussion, and many of them are contained in this book. I would hate to pretend to have all the answers. It is healthy, and it is necessary, to be concerned about the directions our technology is taking us. I am optimistic on most days, worried on others, and mostly afraid only that we may not have the foresight, imagination, and determination we need to do what must be done.

IT IS DIFFICULT to talk about artificial intelligence with anyone for more than a few minutes without crisscrossing between technology, biology, psychology, and philosophy. You can probably add theology and physics in there for good measure, and why not economics and politics too, now that intelligent automation has become vital to business models and its consequences equally important to voters.

In my experience, the tendency of the discussion to expand so rapidly into disparate fields of expertise is frustrating mostly for the technologists. Just about everyone has an opinion on what the technologists are doing, how they are doing it, and what it does and does not mean. The computer people often sound tired of being asked about metaphysical constructs like the mind, let alone the human soul. Meanwhile, programmers and electronics engineers are rarely to be found pestering philosophers and knocking on church doors to discuss the nature of human consciousness, or ringing up politicians to discuss the global security implications of super-intelligent robots. The good news is that at least a few of them do answer the phone when the philosophers and politicians call them.

Many AI researchers do mingle regularly with the neuroscientists and occasionally have stooped to chatting with psychologists, but for the most part they want to be left alone to work on their machines and algorithms in peace. As Ferrucci and Norvig and others have said, they want to solve problems that can be solved, not possibly spend decades investigating things that may have little practical impact even if any progress is made. Life is short and they want to make a difference. The philosophical aspects of AI like "what makes us human?" and "what is intelligence?" can be good for sparking public interest and for attracting the media, but are seen as ephemeral distractions when it comes to getting down to work.

Does it really matter what is or isn't "intelligent" by some definition, no matter how well argued? I concede that the more I learn about it, the less I care. Chess is the perfect example of Larry Tesler's "AI effect," which says that "intelligence is whatever machines haven't done yet." As soon as we figure out a way to get a computer to do something intelligent, like play world championship chess, we decide it's not truly intelligent. Others have pointed out that whenever something becomes practical and common, it stops being called AI at all. It's another illustration that these narratives only matter for a brief point in time.

The exceptions, those who want to tackle the potential of machine cognition by delving into the secrets of human cognition, are often poorly received in the business and academic communities that increasingly prioritize practical results. The largest universities are still something of an exception, but even in the ivy halls and ivory towers there is always a push to publish, to patent, and to profit. The era when giant multinational companies like Bell and government programs like DARPA would pour money into basic research and experimental projects is over. R&D budgets have been slashed over the years as investors take a skeptical view of anything that doesn't feed the bottom line. Government-backed research tends to favor specific gadgets to fit an existing need, not ambitious, open-ended missions to answer big questions like Leonard Kleinrock's "How do we get every computer in the world to talk to each other?"

The Oxford Martin School at Oxford University has collected quite a few of these exceptional people, and also encourages the sort of interdisciplinary associating and free-associating that has gone out of fashion in this era of specialization, benchmarks, and ninety-page grant applications. As a senior visiting fellow there since 2013, I've had the privilege to meet many of these brilliant people, including Nick Bostrom, the author of *Superintelligence*, and other faculty and researchers at his Future of Humanity Institute. Founding Oxford Martin director Ian Goldin thought it would be interesting for me and for his colleagues to have informal workshops where we could talk about the big picture instead of only what was right in front of them in their labs and studies every day.

There's a business saying that if you're the smartest person in the room, you're in the wrong room. Well, after each annual visit to Oxford I could say that it can be tough to feel like the least intelligent person in the room, too. I pride myself on being well informed and generally very good at getting up to speed on complex topics. I read a lot and have plenty of smart friends in different fields who keep me on my mental toes. These Oxford discussions were really on another level, and always ended too soon.

My goal, apart from not sounding like I was the only one in the room without a half-dozen advanced degrees despite that being the case, was to stir the professional pot a little. I asked them to step out of their comfort zones and to talk about what their biggest disappointments were in their fields, and what they thought the public should be paying more attention to. We discussed what the biggest missed predictions were from the previous five years, and then asked them to make a new set for the next five years. I invited them to talk about the bottlenecks in politics and bureaucracy that hold back vital research and the frequently perverse systems for obtaining grants and other funding.

The answers were always fascinating, and it was good to see that these eminent minds were often surprised to hear that their colleagues in a neighboring building were working on something similar, or that they had familiar complaints or concerns. Looking over my notes from the past three years, I was struck by a dilemma that many of them shared, that of working on problems that would help many people today, or on things that would help everyone more in the medium to distant future. Resources are limited, so, as one medical researcher put it, do you work on making better mosquito nets or on a cure for malaria? Of course we can and should try to do both, but it's an illustration of the practical conundrums that even the most vital research faces.

WHAT WAS more important in the long run in my matches against computers? That I might have staved off the inevitable for another few years by being better prepared, or that a machine had achieved the culmination of decades of research and technological advances? I'm sure you will understand that my own answer to this question is a little biased, but I was not going to stand in the way for very long. The 1996–2006 window during which human-machine chess was truly competitive felt like a long time to me because I was on the front line.

From a distance, it's a good example of how human time scales and human capabilities are rendered practically insignificant compared to accelerating technological progress.

If you put this shift on a chart to better understand it, it's easy to see why the spread of AI and automation can be alarming. For centuries, humans were better than machines at chess and everything else requiring cognition. We enjoyed thousands of years of uncontested domination in every intellectual field. Mechanical calculators made a small dent in the nineteenth century, but the real competition only began in the digital age, let's say 1950. From there, it took another forty years for machines to become a serious threat to the top human players, with Deep Thought. Eight years later, I lost to a hugely expensive, custom-designed Deep Blue. Six years after that, better prepared and with more equitable rules, I could only draw two matches against the leading engines, Deep Junior and Deep Fritz, that were at least as strong as Deep Blue despite running on standard servers that only cost a few thousand dollars. In 2006, Vladimir Kramnik, my successor as world champion, lost a match against the latest generation of Fritz with even more favorable regulations by a 4–2 score, ending the age of human-machine play using standard human rules. Any subsequent competitions would require ways of handicapping the machines.

Draw that out as a timeline. Thousands of years of status quo human dominance, a few decades of weak competition, a few years of struggle for supremacy. Then, game over. For the rest of human history, as the timeline draws into infinity, machines will be better than humans at chess. The competition period is a tiny dot on the historical timeline. This is the unavoidable one-way street of technological progress in everything from the cotton gin to manufacturing robots to intelligent agents.

The competition dot gets all the attention because we feel it intensely when it occurs during our lifetimes. The struggle phase often has a direct impact on our lives in real time, so we overinflate its relevance in the big picture. This is not to say it is irrelevant, of course. It is callous to say that all who suffer the consequences of tech disruption should be ignored and just get over it because, in the long run, their suffering won't much matter. The point is that when it comes to looking for solutions to alleviate that suffering, going backwards isn't an option. A corollary is that it is almost always better to start looking for alternatives and how to advance the change into something better instead of trying to fight it and hold on to the dying status quo.

The most important conclusion is not found near the competition dot, but what comes after it, on that long line into eternity. We never go back to the way it was before. No matter how many people are worried about jobs, or the social structure, or killer machines, we can never go back. It's against human progress and against human nature. Once tasks can be done better (cheaper, faster, safer) by machines, humans will only ever do them again for recreation or during power outages. Once technology enables us to do certain things, we never give them up.

POP CULTURE isn't destiny, but I find it significant that tales of the supernatural and medieval fantasy have taken over so much of what used to be the science fiction market. From what can be gleaned from a quick look at the Amazon "science fiction and fantasy" bestseller list, all top twenty books involve vampires, dragons, wizards, or all three. There are many talented authors writing great fantasy stories, and I'm as much a fan of Tolkien and Harry Potter as anyone, but when we look to popular culture for guideposts it is disappointing to see the difficult and valuable work of envisioning the future disposed of with the wave of a wizard's wand.

On the other hand, it would be hard not to have a pessimistic impression of technology after viewing films like James Cameron's *The Terminator* (1984) and the Wachowskis' *The Matrix* (1999). Both stories are based on the theme of man's technology turning against him violently. It's a classic motif, but what makes this old premise more relevant is that since 1980 we have been surrounded by computers, and artificial intelligence is a prominent topic of study and discussion. When the Association for the Advancement of Artificial Intelligence met in

Monterey, California, in 2009, one of the topics its members discussed, and mostly discounted, was the likelihood of humans losing control of super-intelligent computers.

This line of thought, that super-intelligent machines will surpass and possibly turn on their creators, has a long tradition. In a 1951 lecture, Alan Turing suggested that machines would "outstrip our feeble powers" and eventually "take control." Computer scientist and science fiction author Vernor Vinge popularized the concept and coined the modern term for this tipping point, "the singularity," in a 1983 essay. "We will soon create intelligences greater than our own. When this happens, human history will have reached a kind of singularity, an intellectual transition as impenetrable as the knotted space-time at the center of a black hole, and the world will pass far beyond our understanding." A decade later, he added the more specific and menacing lines that are now well known: "Within thirty years, we will have the technological means to create superhuman intelligence. Shortly after, the human era will be ended."

Bostrom picked up that flag and ran with it. He has combined his tremendous range of knowledge with a knack for reaching a mass audience to become an evangelist of the dangers of super-intelligent machines. His book *Superintelligence* goes beyond the usual fearmongering and explains in (still occasionally terrifying) detail the how and why we might create machines that are far more intelligent than we are, and why they might not care to keep humans around anymore.

The prolific inventor and futurist Ray Kurzweil ran in the opposite direction with the concept of super-intelligent machines. His 2005 book, *The Singularity Is Near,* became a bestseller, although, as with so many predictions, "near" is always just close enough to be ominous but never close enough to be in focus. Kurzweil describes a nearly utopian future in which the technological singularity combines genetics and nanotechnology to augment minds and bodies as humans approach an extremely advanced level of cognition and lifespan.

Noel Sharkey has taken a practical approach with his work for establishing ethical norms for autonomous machines, especially "killer robots" in his admirably blunt description. We are very close to those already with drones that do everything but pull the trigger on their own, and the morality and politics of remote killing is something we should be paying attention to already. Sharkey's Foundation for Responsible Robotics also wants us to consider the societal effects of automation as well as its impact on human nature itself. "It is time now to step back and think hard about the future of the technology before it sneaks up and bites us," he says. It's important to have eminent technologists like Sharkey speak out in order to avoid the charge that everyone who wants to pause for a moment is a Luddite fearmonger.

As Sharkey explained to me when we met in Oxford in September 2016, we are on the cusp of a robotics revolution in the workplace—in care, education, sex, transportation, the service industry—as well as in policing and the military. And yet there is a glaring absence of coordinated governmental or international thinking on the topic. He says, "The approach seems to be to just sleepwalk along as we did with the Internet." Sharkey concludes, "Some big figures are out there shouting off their mouths about AI taking over the world and killing us all. I don't see that happening anytime soon. In the meantime, it is kicking up a dust cloud of distraction about the pressing issues of the near future. AI is pretty dumb and narrow despite the hype and yet we are moving towards giving it more control of our lives."

Sharkey's foundation's advocacy for an international bill of human technological rights would define and constrain the kinds of decisions machines can make about humans and human interaction with robots. This immediately brings to mind Asimov's famous "Three Laws of Robotics," but in real life things are far more complex.

When I asked MIT's Andrew McAfee, coauthor of *The Second Machine Age* and *Race Against the Machine*, what he thought was the biggest misunderstanding about artificial intelligence today, he was succinct: "The greatest misconception is the hope that the singularity—or the fear that super-intelligence—is right around the corner." McAfee's commonsensical and humane investigations into the impact of technology on society most closely match my own outlook. His pragmatism matches the great line by machine learning expert Andrew Ng, formerly of Google and now with China's Baidu, who has said that worrying about super-intelligent and evil AI today is like worrying about "the problem of overcrowding on Mars."

This is not to say I'm not grateful that there are people like Bostrom worrying about these things. I just want them to do most of the worrying for me since there are so many immediate issues to deal with in the meanwhile. I am prone to seeing even clearly harmful side effects as growing pains that will turn out to be far less consequential than they may appear in the early days of a new technology. New isn't always better, but it's just as wrong to believe that new is always worse, and a pessimistic view is more detrimental to the development of our civilization.

We cannot be sure what changes our new technology will bring, but I trust the young people who are growing up with it. I trust that they will find surprising new ways to use technology the way my generation used computers and satellites and how every generation has used technology to fulfill human ambition.

CONCLUSIONS ARE usually for winding down, but I would prefer to use this one to stir things up. I hope you will take this section as a reading list and as an invitation to take an active role in creating the future you want to see. This debate is unique because it is not academic. It is not a postmortem. The more that people believe in a positive future for technology, the greater chance there is of having one. We will all choose what the future looks like by our beliefs and our actions. I do not believe in fates beyond our control. Nothing is decided. None of us are spectators. The game is under way and we are all on the board. The only way to win is to think bigger and to think deeper.

This is not a choice between utopia or dystopia. It is not a matter of us versus anything else. We will need every bit of our ambition in order to stay ahead of our technology. We are fantastic at teaching our machines how to do our tasks, and we will only get better at it. The only solution is to keep creating new tasks, new missions, new industries that even we don't know how to do ourselves. We need new frontiers and the will to explore them. Our technology excels at removing the difficulty and uncertainty from our lives, and so we must seek out ever more difficult and uncertain challenges.

I have argued that our technology can make us more human by freeing us to be more creative, but there is more to being human than creativity. We have other qualities the machines cannot match. They have instructions while we have purpose. Machines cannot dream, not even in sleep mode. Humans can, and we will need our intelligent machines in order to turn our grandest dreams into reality. If we stop dreaming big dreams, if we stop looking for a greater purpose, then we may as well be machines ourselves.

ACKNOWLEDGMENTS

I would like to thank the many pioneers from the worlds of chess and computer science who worked together on what became the longest-running science experiment in history: the quest to build a world-champion chess machine. My life and my career were made immeasurably richer by standing on their shoulders and by participating in this quest. Many of their names and contributions are highlighted throughout this book, while several have been my worthy opponents and others my fast friends.

Frederic Friedel sparked my interest in chess machines, although he loves them so much I've never been completely sure whose side he's on. Ken Thompson lent his invaluable expertise and goodwill to my sport and to my matches against machines. David Levy and Monty Newborn saw computer chess as a way to teach the world about machine intelligence and chess. Jonathan Schaeffer, Anthony Marsland, and Donald Michie contributed decades of insightful writing on game-playing machines in addition to their many technical achievements. Matthias Wüllenweber and Frans Morsch created ChessBase and Fritz, the programs that defined the computer era of professional chess. Thomas Anantharaman, Murray Campbell, Joseph Hoane, and Feng-hsiung Hsu created Deep Thought at Carnegie Mellon, which turned into Deep Blue at IBM. They deservedly seized the grail dreamt of by Alan Turing, Claude Shannon, and Norbert Wiener, and it was my fortune, not misfortune, to be holding it at the time. My friend Shay Bushinsky and his colleague Amir Ban created the remarkable program Junior, my opponent in my final human-machine match in 2003.

In recent years, many experts have had the patience to personally contribute to my education in artificial intelligence and robotics. Nick

Bostrom and his colleagues at Oxford Martin's Future of Humanity Institute; Andrew McAfee at MIT; Noel Sharkey at the University of Sheffield; Nigel Crook at Oxford Brookes University; David Ferrucci at Bridgewater. I've never met Douglas Hofstadter or Hans Moravec, but their writings on human and machine cognition are especially provocative and essential.

Special thanks to: My agent at the Gernert Company, Chris Parris-Lamb, and my editor at PublicAffairs, Ben Adams. They have shown impressive resilience in adjusting deadlines to accommodate a chess-player's eternal love of time trouble. Peter Osnos, Clive Priddle, and Jaime Leifer, the terrific team at PublicAffairs. My collaborator of nearly nineteen years, Mig Greengard, whose former lives in programming and chess made him even more indispensable than usual on this project.

NOTES

INTRODUCTION

2 *"It is comparatively easy to make computers exhibit adult level performance."* Hans Moravec, *Mind Children* (Cambridge, MA: Harvard University Press, 1988).

5 *Deep Blue matches beyond what was publicly known.* A notable exception was the 2003 documentary film about the match, *Game Over: Kasparov and the Machine.* But while it succeeded in reflecting my perspective it was content to leave much to conjecture. This made for good drama and cinema, but it lacked the rigor and depth I finally felt ready to apply in this book.

8 *According to the Associated Press, "Thousands struggled up stairways."* Associated Press, September 24, 1945. Online via the *Tuscaloosa News*: https://news.google.com/newspapers?nid=1817&dat=19450924&id=I-4-AAAAIBAJ&sjid=HEoMAAAAIBAJ&pg=4761,2420304&hl=en. On a related note, the impact of technology on the age-old battle between labor and capital is critical for any discussion on rising economic inequality.

CHAPTER 1. THE BRAIN GAME

12 *The game is popular on every continent.* Chess did not only move westward, it also spread east, where its forms took on distinctive cultural flavors. Many East Asian countries have their own chess variants, likely also descended from an Indian precursor, that are more popular there than modern "European" chess. Japan has shogi, China has xiangqi, and much of the region is also devoted to Go, which is unrelated to chess and is even older.

14 *A character of Goethe's called chess a "touchstone of the intellect,"* The character Adelheid calls chess "a touchstone of the intellect" in Goethe's 1773 drama, *Götz von Berlichingen.*

15 *"The willingness to take on new challenges."* The *Der Spiegel* article titled "Genius and Blackouts" was published in issue 52 in 1987, in German here: http://www.spiegel.de/spiegel/print/d-13526693.html.

16 *"a phenomenon in the history of man."* Cited in H. J. R. Murray's *A History of Chess* as appearing in an article in the *World* newspaper on May 28, 1782.

16 *set by a German player of average master strength.* Marc Lang is a German FIDE master rated around 2300. He played forty-six boards blindfold in 2011. The old records were often controversial because the conditions were not standardized. For example, some players had access to the scoresheets of the games. More on Lang's record at https://www.theguardian.com/sport/2011/dec/30/chess-marc.

19 *military exemptions were given to strong chessplayers.* I. Z. Romanov, *Petr Romanovskii* (Moscow: Fizkultura i sport, 1984), 27.

19 *and the Communist system that produced him.* Typical of Stalin's cult of personality, a game was published in which he supposedly defeated Nikolai Yezhov, the future head of the secret police, in elegant fashion.

19 *winning the gold medal in eighteen of the nineteen Chess Olympiads.* Hungary relegated the USSR to silver in 1978, considered a huge humiliation. When I was just seventeen I was a part of the "comeback team" that won gold in 1980.

20 *proudly exchange my Soviet flag for a Russian one hastily handmade by my mother, Klara.* I insisted on changing flags over the protest of Soviet sports officials and my opponent, Karpov. For the full story, see my 2015 book, *Winter Is Coming.*

CHAPTER 2. RISE OF THE CHESS MACHINES

28 *In "Programming a Computer for Playing Chess."* Claude Shannon, "Programming a Computer for Playing Chess," *Philosophical Magazine* 41, ser. 7, no. 314, March 1950. It was first presented at the National Institute of Radio Engineers Convention, March 9, 1949, New York.

29 *This insight echoes Norbert Wiener's note.* Norbert Wiener, *Cybernetics or Control and Communication in Animal and Machine* (New York, Technology Press, 1948), 193.

33 *made an accurately calculated piece sacrifice.* Mikhail Tal, *The Life and Games of Mikhail Tal* (London: RHM, 1976), 64.

34 *enough to beat a very weak human player.* This was indeed a very optimistic number, and a chess machine wouldn't reach the speed of analyzing a million moves per second until the 1990s. But long before that happened, efficient algorithms had made pure Type A programs obsolete.

36 *roughly in accord with Moore's Law.* Moore's law, popularly understood to say that computing power will double every two years, has been a golden rule of technology for decades. As with so many popular maxims, Gordon Moore's original statement was more specific and was later updated by

him. In 1965, Moore, the cofounder of Intel, referred to how the density of transistors on integrated circuits had doubled every year since they had been invented. In 1975, he updated his prediction to every two years.

37 *better than Deep Blue did in 1997 on its specialized hardware.* For additional perspective on the practical implications of Moore's law and how rapidly computers have gotten faster and smaller, the 1985 Cray-2, again the world's fastest computer at the time, weighed several thousand pounds and had a peak speed of 1.9 gigaflops while the 2016 iPhone 7 weighs five ounces and reaches 172 gigaflops.

CHAPTER 3. HUMAN VERSUS MACHINE

47 *Many things on Earth are faster than Usain Bolt's top speed.* Legendary American gold medalist Jesse Owens, hero of the 1936 Berlin Olympic Games, actually did run stunt races against horses, dogs, cars, and motorcycles in the 1940s.

48 *The most popular programs were directed toward casual consumers.* The slogan for a PC game called Battle Chess, which appeared in 1988: "It took 2,000 years for someone to make chess better!" I think not.

52 *Bronstein proposed many innovative ideas for promoting chess.* Bronstein suggested shuffling the pieces for each game long before Bobby Fischer proposed a version of doing this that is now fairly popular. Also in advance of Fischer, Bronstein proposed a time delay for each move to ensure that the players would always have at least a few seconds to move. Time delay or increment is now standard in professional events.

52 *In 1963, Bronstein was still one of the strongest players in the world.* There have always been allegations that Bronstein was not "allowed" to beat Botvinnik, a loyal Soviet man, an echo of my confrontations with Karpov decades later.

52 *The basic set of values was established two centuries ago.* Different players, like different computer programs, have proposed slight modifications in the piece values. The most radical was probably Bobby Fischer, who suggested bishops were worth 3.25 pawns.

56 *"the AGAT wouldn't stand a chance in today's international market,"* Leo D. Bores, "AGAT: A Soviet Apple II Computer," *BYTE* 9, no. 12 (November 1984).

58 *I conceded defeat to avoid having to sit watching through dinnertime.* A version of this anecdote appears in *How Life Imitates Chess*. In the ten years since I wrote that book, it has become even clearer to me that technology is like language, best learned through early immersion.

58 *It was a much-coveted type of hard drive.* If I recall correctly, the shouting was being done by Stepan Pachikov, a computer scientist who shared the

direction of the computer club with me. His contributions to handwriting recognition software at the Soviet company ParaGraph were used in the Apple Newton. He later moved to Silicon Valley and founded Evernote, the ubiquitous note-taking app.

61 *I once made a television commercial for the search engine company AltaVista.* If you want to know what happened to AltaVista, you can google it!

65 *This fits the axiom of Bill Gates.* Bill Gates, *The Road Ahead* (New York: Viking Penguin, 1995).

CHAPTER 4. WHAT MATTERS TO A MACHINE?

69 *"I checked it very thoroughly," said the computer.* Douglas Adams, *The Hitchhiker's Guide to the Galaxy* (New York: Del Rey, 1995), Kindle edition, locations 2606–14.

69 *In my lectures on the human-machine relationship, I'm fond of citing Pablo Picasso.* Different versions of this are cited in William Fifield's original interview with Picasso, "Pablo Picasso: A Composite Interview," published in the *Paris Review* 32, Summer–Fall 1964, and in Fifield's 1982 book, *In Search of Genius* (New York: William Morrow).

70 *They believed it was worthwhile to fund Ferrucci's attempts.* Steve Lohr, "David Ferrucci: Life After Watson," *New York Times*, May 6, 2013.

73 *In a 1989 article, two of the leading figures in computer chess wrote an essay.* Mikhail Donskoy and Jonathan Schaeffer, "Perspectives on Falling from Grace," *Journal of the International Computer Chess Association* 12, no. 3, 155–63.

74 *"But one is born an excellent player."* Binet's conclusions about chess players are from several of his papers from 1893 and are usefully summarized in the book *A Century of Contributions to Gifted Education: Illuminating Lives* by Ann Robinson and Jennifer Jolly (New York and London: Routledge, 2013).

74 *John McCarthy, the American computer scientist who coined the term "artificial intelligence" in 1956.* McCarthy later credited the *Drosophila* phrase to his Soviet peer Alexander Kronrod.

CHAPTER 5. WHAT MAKES A MIND

79 *I am not going to argue with the International Olympic Committee.* I don't doubt that if a mind sport proved lucrative enough, the International Olympic Committee would quickly change its tune about the definition of physical exertion. But here bridge has an advantage over chess and video games (e-sports) have an advantage over both.

82 *"natural ability requires a huge investment of time in order to be made manifest."* Malcolm Gladwell post on Reddit, https://www.reddit.com/r/IAmA

/comments/274oct/hi_im_malcolm_gladwell_author_of_the_tipping/chx6dpv/.

85 *Might I have become a Shogi champion had I been born in Japan.* In my visit to Tokyo in 2014 to promote a human-machine shogi competition we joked that in Japan I was the "Habu of Western chess." High praise!

85 *I would hate to provide anyone with a genetic excuse for taking it easy.* Several recent studies have indicated that practice is indeed substantially heritable. This isn't exactly what I meant when I first wrote "hard work is a talent" in 2007, but it's always nice to see scientific research confirm your assumptions. See https://www.ncbi.nlm.nih.gov/pubmed/24957535 and http://pss.sagepub.com/content/25/9/1795 for research using thousands of pairs of twins to measure the heritability of work ethic.

91 *What has this to do with the skill.* Donald Michie, "Brute Force in Chess and Science," collected in *Computers, Chess, and Cognition* (Berlin: Springer-Verlag, 1990).

92 *Fischer answered, "How would you know?"* I was told this story in Buenos Aires, Argentina, and have no way to know if it's true. But it definitely sounds like something Fischer might say. It is also bitingly insightful, as few fans would have any idea of the quality of a world champion's game without expert commentary. Today it's quite different, when everyone has a super-strong engine at his disposal and feels empowered to scoff at the champion's mistakes as if they'd found them themselves.

CHAPTER 6. INTO THE ARENA

95 *within five to ten years that some of these tough problems would be solved."* Remarks by Bill Gates, International Joint Conference on Artificial Intelligence, Seattle, Washington, August 7, 2001, https://web.archive.org/web/20070515093349/http://www.microsoft.com/presspass/exec/billg/speeches/2001/08-07aiconference.aspx.

99 *DARPA has proposed tournament competitions.* Including a proposal to develop "Deep Capture the Flag." See https://cgc.darpa.mil/Competitor_Day_CGC_Presentation_distar_21978.pdf.

100 *"Data trumps everything."* Josh Estelle, quoted in the *Atlantic*, November 2013, "The Man Who Would Teach Machines to Think," by James Somers.

100 *educated on a diet of GM games, giving up its queen was clearly the key.* Recounted by Kathleen Spracklen, the creator of the famous microcomputer program Sargon, along with her husband, Dan. "Oral History of Kathleen and Dan Spracklen," interview by Gardner Hendrie, March 2, 2005, http://archive.computerhistory.org/projects/chess/related_materials/oral-history/spacklen.oral_history.2005.102630821/spracklen.oral_history_transcript.2005.102630821.pdf.

101 *Watson then answered simply "leg."* It was Watson's first night on the show. You can watch the "leg" clip online, and it's also amusing to see the many YouTube comments from humans (one assumes) who were delighted by the machine's failures. Don't make them angry! *Jeopardy*, aired February 14, 2011, https://www.youtube.com/watch?v=fJFtNp2FzdQ.

102 *A "lounge for the weak" at an airport, a "plate of little stupids" at a restaurant.* Weak instead of tired, so a rest area. The second one makes perfect sense if you know that (1) a burrito is Mexican food; (2) *burro* is also Mexican slang for stupid; (3) the *-ito* suffix is a diminutive in Spanish. Burritos = little burros = little stupids.

104 *I don't know why more people aren't that way."* James Somers, "The Man Who Would Teach Machines to Think," *Atlantic*, November 2013.

108 *the reason the project was started in the first place."* F-h. Hsu, T. S. Anantharaman, M. S. Campbell, and A. Nowatzyk, "Deep Thought," in *Computers, Chess, and Cognition*, Schaeffer and Marsland, eds. (New York: Springer-Verlag, 1990).

108 *as a competitive sport (performance driven) rather than as a science (problem driven).* Danny Kopec, "Advances in Man-Machine Play," in *Computers, Chess, and Cognition*, Schaeffer and Marsland, eds. (New York: Springer-Verlag, 1990).

109 *There just weren't any women on the horizon who showed the potential.* I won't hide from the fact that I did make regrettably sexist remarks about women in chess around this time. In that 1989 *Playboy* interview I said men were better at chess because "women are weaker fighters" and that "probably the answer is in the genes." The possibility of gender brain differences aside, I find it almost hard to believe I said this considering that my mother is the toughest fighter I know.

110 *I'm a little chagrined now to see that I did not play the best moves throughout.* If you are interested—43.Qb1—a clever move I don't see mentioned in any of the many books and articles that covered the match. Black is still much better but it will take a lot of work to break through. I could have kept my crushing advantage with 40..f5. The free chess engine on my laptop finds 43.Qb1 in half a second, to indicate how far things have come.

110 *an advantage similar to that of serving in tennis.* I don't mean this strictly statistically, since serving in tennis confers a far greater advantage than having the white pieces in chess. But it's similar in how both confer the initiative, the ability to better control the development of the game.

111 *wrote the* New York Post, *with an anachronistic Cold War jab.* Andrea Privitere, "Red Chess King Quick Fries Deep Thought's Chips," *New York Post*, October 23, 1989.

CHAPTER 7. INTO THE DEEP END

114 *"Beating Gary Kasparov at chess is considerably more difficult than climbing Mount Everest."* Raymond Keene, *How to Beat Gary Kasparov at Chess* (New York: Macmillan, 1990). Publications deciding on the English spelling of my first name used to fluctuate between Gary, Garry, and even Garri, but I prefer Garry.

119 *Privacy is dying, so transparency must increase.* For a look at how society might cope in a post-privacy world, I recommend David Brin's 1997 book, *The Transparent Society*, and the updates and conversations about it on his website.

120 *"the best commercial chess programs appear to have measurably better evaluation than the research."* Hsu et al., "Deep Thought," in *Computers, Chess, and Cognition.*

124 *I had to agree to a draw. I was out.* In the next round, Genius beat GM Predrag Nikolic and was then beaten in the semifinal by Viswanathan Anand.

129 *"It won about nine out of ten games against Fritz."* Feng-hsiung Hsu, *Behind Deep Blue* (Princeton, NJ: Princeton University Press, 2002).

131 *Deep Blue committed suicide.* The "reboot induced" mistake was 13.0-0 , instead of the stronger 13.g3, which is what one observer said Deep Blue was planning on playing before the disconnection. Then it blundered with 14.Kh1 only to get a reprieve when Fritz missed 14..Bg4, winning immediately. Two moves later, 16.c4 was the losing blunder, punished immediately by 16..Qh4, and there was never a chance after that for white to save the game. Curiously, Hsu highlighted the 16.c4 blunder a few days after the match in a post on an online chess discussion group but omitted it in his book.

136 *they wouldn't expect me to repeat that game.* I know that technically the machine I faced in 1989 was Deep Thought not Deep Blue and that it was practically a different machine altogether, but if only for convenience I will always consider the 1989, 1996, and 1997 matches to be against different iterations of the same opponent.

138 *less accurate than when they were playing.* I give a specific historical example of this in *How Life Imitates Chess*, a world championship game between Lasker and Steinitz from 1894 that had been misrepresented mightily for over a century.

140 *"not allowing your opponent to show you what he can do."* Brad Leithauser, "Kasparov Beats Deep Thought," *New York Times*, January 14, 1990.

140 *I might have saved the game.* By playing 27..f4 immediately instead of the error 27..d4. 27..Rd8 was also okay for black.

141 *no human can be sure to have seen everything. Deep Blue can.* Charles Krauthammer, "Deep Blue Funk," *TIME,* June 24, 2001.

141 *I could sense "a new kind of intelligence across the table."* Garry Kasparov, "The Day I Sensed a New Kind of Intelligence," *TIME,* March 25, 1996.

149 *the rest of the Dow Jones went down significantly.* Of course there is no way to prove the match was responsible for this, but, as Newborn points out, even if only 10 percent of the rise was due to the match, that's over $300 million in value. Not bad for six games of machine chess.

CHAPTER 8. DEEPER BLUE

155 *Imagine what winning a match might do.* Or, as Hsu puts it in his book *Behind Deep Blue,* "The event could only get bigger. There was no way in hell that IBM would not want a rematch."

157 *Botvinnik dominated the rematch.* Tal's health, never good, was quite poor during parts of the rematch, but it was also apparent that Botvinnik had come very well prepared.

158 *"Conceit does not put one in the right frame of mind for work."* Mikhail Botvinnik, *Achieving the Aim* (Oxford, UK: Pergamon Press, 1981), 149. The quote is from the English translation of his book, first published in Russian in 1978.

158 *"he stumbled on something that he was able to exploit."* Monty Newborn, *Deep Blue: An Artificial Intelligence Milestone* (New York: Springer-Verlag, 2003), 103.

158 *As was later revealed in Michael Khodarkovsky's book.* Michael Khodarkovsky and Leonid Shamkovich, *A New Era* (New York: Ballantine, 1997).

161 *"This time, we're just going to play chess."* Bruce Weber, "Chess Computer Seeking Revenge Against Kasparov," *New York Times,* August 20, 1996.

162 *C. J. Tan and others still occasionally referred to future cooperation with me.* The Club Kasparov website did launch in beta form right before the match, but the plug was pulled on it almost as quickly as it was on Deep Blue itself. I personally supported it in Russia and, in 1999, it was relaunched as Kasparov Chess Online with new venture capital.

CHAPTER 9. THE BOARD IS IN FLAMES!

177 *"'Here I was blind, I didn't see this!'"* Dirk Jan ten Geuzendam, "I Like to Play with the Hands," *New In Chess,* July 1988, 36–42.

179 *each story containing more errors about chess.* The *Wired*'s "Did a Computer Bug Help Deep Blue Beat Kasparov?" story from September 28, 2012, by Klint Finley, deserves to be singled out because it mixes everything up so spectacularly it could have been written by a computer. It

confuses the rook move blunder from game one with Deep Blue's bishop move in game two, and by so doing gives the credit for Deep Blue's most remarkable maneuver to a random bug.

184 *C. J. Tan's pre-rematch statement that "the science experiment is over."* Robert Byrne, "In Late Flourish, a Human Outcalculates a Calculator," *New York Times*, May 4, 1997.

184 *his work on Deep Blue and other events.* Dirk Jan ten Geuzendam, "Interview with Miguel Illescas," *New In Chess*, May 2009.

196 *"a little with the hand of God."* Later in the game Maradona would help everyone but the English forget *la mano de Dios* goal by scoring the sensational "goal of the century" after running past half the English team.

CHAPTER 10. THE HOLY GRAIL

202 *"because IBM had insisted he sign a secrecy agreement."* Bruce Weber, "Deep Blue Escapes with Draw to Force Decisive Last Game," *New York Times*, May 11, 1997.

204 *I missed one good attacking chance.* My last best chance to win the game was likely 35..Rff2. Incredibly, after my 35..Rxg4 there appears to be no clear win for black.

208 *they had specially requisitioned from an expert.* Murray Campbell, A. Joseph Hoane Jr., and Feng-hsiung Hsu, "Deep Blue," *Artificial Intelligence* 134, 2002, 57–83.

212 *I did miss a win in the game five endgame.* In game five 44.Rd7 is winning, instead of my 44.Nf4. Deep Blue blundered with 43..Nd2 when 43..Rg2 draws.

217 *As Pynchon's "Proverbs for Paranoids, 3" says.* Thomas Pynchon, *Gravity's Rainbow* (New York: Viking, 1973), 251. Here are all five of the Proverbs for Paranoids, several of which seem disturbingly applicable here, although I won't say which ones. "1. You may never get to touch the Master, but you can tickle his creatures. 2. The innocence of the creature is in inverse proportion to the immorality of the Master. 3. If they can get you asking the wrong questions, they don't have to worry about answers. 4. You hide, They seek. 5. Paranoids are not paranoids because they're paranoid, but because they keep putting themselves, fucking idiots, deliberately into paranoid situations."

CHAPTER 11. HUMAN PLUS MACHINE

222 *but on our creation and use of tools.* The works of cognitive scientist Steven Pinker and his colleagues has convinced me that the origins of the development of human language are unknown and possibly unknowable,

as befits "the hardest problem in science," as Pinker's essay on the subject is titled. It was probably fortuitous that I did not have the chance to discuss language evolution with him during our brief encounters at the Oslo Freedom Forum, or this book might have ended up being even longer. And so, I will stay with tools and other things that can be verified by archeologists. And the ability to speak beyond rudimentary sounds wasn't going to save cave dwellers from freezing or starving. Furs, fire, and spears would. See Morten H. Christiansen and Simon Kirby, eds. *Language Evolution: The Hardest Problem in Science?* (New York: Oxford University Press, 2003).

223 *Cory Doctorow coined the term "outboard brain."* Cory Doctorow, "My Blog, My Outboard Brain," May 31, 2002, http://archive.oreilly.com/pub/a/javascript/2002/01/01/cory.html.

223 *"we've outsourced important peripheral brain functions to the silicon."* Clive Thompson, "Your Outboard Brain Knows All," *Wired*, September 25, 2007.

224 *"It's merely my autonomy that I'm losing."* David Brooks, "The Outsourced Brain," *New York Times*, October 26, 2007. His tone is derisive here, or at least resigned, although Brooks has in the past been an accurate chronicler of American cultural foibles. His book *Bobos in Paradise* describes the search for fake authenticity by the entitled, and a similar attitude decries the new technology we need for supplanting an obsolete analog past.

225 *"Does an overreliance on machine memory shut down other important ways."* Thompson, "Your Outboard Brain Knows All."

231 *more wins and losses than draws.* Since Kramnik's use of the Berlin Defense in our 2000 world championship match brought it to prominence, 63 percent of the elite games in which it appears have been drawn. Compare this to my old favorite, the Sicilian Defense, which was drawn 49 percent of the time over the same period.

238 *to monitor and control one's mood is of great importance.* Patrick Wolff, *Kasparov versus Anand* (Cambridge: H3 Publications, 1996).

239 *decision making that is slower, more conservative, and inferior.* This 2011 study is a good overview: "Decision-Making and Depressive Symptomatology" by Yan Leykin, Carolyn Sewell Roberts, and Robert J. DeRubeis, https://www.ncbi.nlm.nih.gov/pmc/articles/PMC3132433/.

239 *"potential disappointment in the expected outcome."* Wolff, *Kasparov versus Anand.*

239 *Depression short-circuits intuition.* There are many studies on this topic as well. An interesting and recent one is discussed on the website of the British Psychological Society: "When we get depressed, we lose our

ability to go with our gut instincts," https://digest.bps.org.uk/2014/11/07/when-we-get-depressed-we-lose-our-ability-to-go-with-our-gut-instincts/.

248 *enhance human decisions instead of replacing them.* Murray Campbell of the Deep Blue team is one of the leaders on the IA project at IBM. Does that mean he's come over to my side?!

CONCLUSION. ONWARD AND UPWARD

250 *"A man is much more dispensable than a computer."* Isaac Asimov, "The Feeling of Power" in *If*, February 1958.

252 *what was right in front of them in their labs and studies.* Ian Goldin wrote an important book, *Age of Discovery: Navigating the Risks and Rewards of Our New Renaissance*, and left Oxford Martin in mid-2016. The new director is Achim Steiner.

256 *"the world will pass far beyond our understanding."* Vernor Vinge in an op-ed in *Omni* magazine, January 1983.

256 *"we will have the technological means to create superhuman intelligence."* Vernor Vinge, "The Coming Technological Singularity: How to Survive in the Post-Human Era," originally in *Vision-21: Interdisciplinary Science and Engineering in the Era of Cyberspace*, G. A. Landis, ed., NASA Publication CP-10129, 11–22, 1993.

257 *in real life things are far more complex.* Asimov's Three Laws of Robotics: "A robot may not injure a human being or, through inaction, allow a human being to come to harm. A robot must obey the orders given it by human beings except where such orders would conflict with the First Law. A robot must protect its own existence as long as such protection does not conflict with the First or Second Laws." Isaac Asimov, *I, Robot* (New York: Gnome Press, 1950).

INDEX

ABOUT THE AUTHORS

Igor Khodzinskiy, 2013

Garry Kasparov is a business speaker, global human rights activist, author, and former world chess champion. His keynote lectures and seminars on strategic thinking, achieving peak performance, and tech innovation have been acclaimed in dozens of countries. A frequent contributor to the *Wall Street Journal* and dozens of other publications, he is the author of two books, *How Life Imitates Chess* and *Winter Is Coming,* each of which has been translated into more than a dozen languages. He is a member of the executive advisory board of the Foundation for Responsible Robotics and a senior visiting fellow at the Oxford Martin School, with a focus on interdisciplinary research and human-machine decision making.

Mig Greengard has been Garry Kasparov's spokesman and adviser since 1998. Their collaboration spans hundreds of articles, speeches, and the books *How Life Imitates Chess* and *Winter Is Coming.* He lives in Brooklyn.

PublicAffairs is a publishing house founded in 1997. It is a tribute to the standards, values, and flair of three persons who have served as mentors to countless reporters, writers, editors, and book people of all kinds, including me.

I. F. Stone, proprietor of *I. F. Stone's Weekly*, combined a commitment to the First Amendment with entrepreneurial zeal and reporting skill and became one of the great independent journalists in American history. At the age of eighty, Izzy published *The Trial of Socrates*, which was a national bestseller. He wrote the book after he taught himself ancient Greek.

Benjamin C. Bradlee was for nearly thirty years the charismatic editorial leader of *The Washington Post*. It was Ben who gave the *Post* the range and courage to pursue such historic issues as Watergate. He supported his reporters with a tenacity that made them fearless and it is no accident that so many became authors of influential, best-selling books.

Robert L. Bernstein, the chief executive of Random House for more than a quarter century, guided one of the nation's premier publishing houses. Bob was personally responsible for many books of political dissent and argument that challenged tyranny around the globe. He is also the founder and longtime chair of Human Rights Watch, one of the most respected human rights organizations in the world.

• • •

For fifty years, the banner of Public Affairs Press was carried by its owner Morris B. Schnapper, who published Gandhi, Nasser, Toynbee, Truman, and about 1,500 other authors. In 1983, Schnapper was described by *The Washington Post* as "a redoubtable gadfly." His legacy will endure in the books to come.

Peter Osnos, *Founder and Editor-at-Large*